RAND NATIONAL SECURITY RESEARCH DIVISION

A Surprise Out of Zion?

Case Studies in Israel's Decisions on Whether to Alert the United States to Preemptive and Preventive Strikes, from Suez to the Syrian Nuclear Reactor

Warren Bass

Approved for public release; distribution unlimited

For more information on this publication, visit www.rand.org/t/RR498

Library of Congress Cataloging-in-Publication Data is available for this publication.

ISBN: 978-0-8330-8416-3

Published by the RAND Corporation, Santa Monica, Calif.
© Copyright 2015 RAND Corporation

RAND® is a registered trademark.

Limited Print and Electronic Distribution Rights

This document and trademark(s) contained herein are protected by law. This representation of RAND intellectual property is provided for noncommercial use only. Unauthorized posting of this publication online is prohibited. Permission is given to duplicate this document for personal use only, as long as it is unaltered and complete. Permission is required from RAND to reproduce, or reuse in another form, any of its research documents for commercial use. For information on reprint and linking permissions, please visit www.rand.org/pubs/permissions.html.

The RAND Corporation is a research organization that develops solutions to public policy challenges to help make communities throughout the world safer and more secure, healthier and more prosperous. RAND is nonprofit, nonpartisan, and committed to the public interest.

RAND's publications do not necessarily reflect the opinions of its research clients and sponsors.

Support RAND

Make a tax-deductible charitable contribution at
www.rand.org/giving/contribute

www.rand.org

Preface and Summary

Might senior U.S. policymakers be surprised by an Israeli strike on Iran's nuclear facilities? This study considers four key historical precedents to shed some light on today's decisionmaking in both the United States and Israel. In 1956, the Eisenhower administration was livid over being surprised by Israel's intervention in Egypt; in 1967, Israel repeatedly urged the Johnson administration to approve its use of force during the crisis that led to the Six-Day War; in 1981, the Reagan administration was startled by Israel's strike on Iraq's nuclear reactor; and in 2007, the George W. Bush administration ultimately rejected Israel's high-level requests for a U.S. bombing campaign against a Syrian nuclear facility.

This study seeks to use historical narrative to inform the reader's understanding of choices both past and present, over several decades in which the U.S.-Israel relationship has grown far closer and deeper. For these purposes, we may think of Israeli leaders as falling into two categories: confronters and consulters. Israeli Prime Ministers David Ben-Gurion and Menachem Begin presented the United States with *faits accomplis* in 1956 and 1981, running serious risks in the bilateral relationship; by contrast, Levi Eshkol and Ehud Olmert took pains to try to see if Washington would resolve Israel's security dilemmas in 1967 and 2007. In neither instance did consultation result in a U.S. use of force on Israel's behalf, but in both cases, it did yield considerable dividends of U.S. understanding when Israel ultimately took matters into its own hands. From Suez on, one thing has not changed: Superpowers do not like being surprised.

This research was conducted within the International Security and Defense Policy Center of the RAND National Defense Research Institute, a federally funded research and development center sponsored by the Office of the Secretary of Defense, the Joint Staff, the Unified Combatant Commands, the Navy, the Marine Corps, the defense agencies, and the defense Intelligence Community.

For more information on the International Security and Defense Policy Center, see http://www.rand.org/nsrd/ndri/centers/isdp.html or contact the director (contact information is provided on the web page).

Contents

Preface and Summary ... iii
Acknowledgments ... vii
Abbreviations ... ix

Introduction ... 1
Case I: The Suez War, 1956 ... 5
Case II: The Six-Day War, 1967 ... 15
Case III: The Raid on Osiraq, 1981 ... 27
Case IV: The Raid on al-Kibar, 2007 ... 45
Conclusion ... 61

References ... 75

Acknowledgments

Many thanks to Olya Oliker for her help and support in launching this project. Jim Dobbins, Seth Jones, Eric Peltz, Robin Meili, Ali Nader, Andrew Weiss, and many other RAND colleagues offered valuable input and insight along the way.

Ambassador Sam Lewis generously sat for a candid and fascinating interview about Osiraq. Rami Shtivi of the Menachem Begin Heritage Center in Jerusalem kindly walked me through some of its archival holdings.

David Makovsky of the Washington Institute for Near East Policy and Dalia Dassa Kaye of RAND both read the paper carefully and offered typically thoughtful and helpful comments.

I'm also particularly grateful to Joan Chanman-Forbes for her help with funding and to Maggie Snyder for stalwart assistance throughout.

Abbreviations

AWACS	Airborne Warning and Control System
CIA	Central Intelligence Agency
DNI	Director of National Intelligence
DOD	Department of Defense
IAEA	International Atomic Energy Agency
IDF	Israel Defense Forces
NATO	North Atlantic Treaty Organization
NGA	National Geospatial-Intelligence Agency
NSC	National Security Council
UN	United Nations
UNEF	United Nations Emergency Force
WMD	weapons of mass destruction

Introduction

For years now, the United States and Israel have held close consultations about Iran's nuclear ambitions. Much ink has already been spilled about whether Israel will ultimately choose to attack Iran. But far less attention has been paid to the question of whether U.S. officials are likely to be *surprised* by any Israeli strike. This study seeks to offer some perspective on that question by considering four key cases in which Israeli prime ministers were faced with thorny questions of whether to notify or consult with the United States over looming preemptive or preventive military strikes.[1]

[1] When it comes to the anticipatory use of military force, the terms "preemptive" and "preventive" are often used interchangeably. They should not be. As Michael Walzer has put it, preventive war "responds to a distant danger, a matter of foresight and free choice," usually to shore up a regional balance of power or avert a growing threat. In describing preemptive force, on the other hand, Walzer cites Secretary of State Daniel Webster from 1842, who argued that preemptive force could only be justified by the urgent and immediate need to defend oneself from a blow about to be struck, from an attack that can be seen coming but has yet to land—"instant, overwhelming, leaving no choice of means, and no moment for deliberation." Walzer himself draws a somewhat different definition of a legitimate act of preemption in which a state faces "a manifest intent to injure, a degree of active preparation that makes that intent a positive danger, and a general situation in which waiting, or doing anything other than fighting, greatly magnifies the risk." To focus on the key question at hand, this study has limited its scope to cases in which Israel did strike preventively or preemptively—not such fascinating but different cases as the Yom Kippur War of 1973, in which Israel chose with a heavy heart to absorb the first blow. See Michael Walzer, *Just and Unjust Wars: A Moral Argument With Historical Illustrations*, New York: Basic, 2006, pp. 74–75, 80–81. See also Karl Mueller et al., *Striking First: Preemptive and Preventive Attack in U.S. National Security Policy*, Santa Monica, Calif.: RAND Corporation, MG-403-AF, 2006.

This paper seeks to offer a different angle on current decisionmaking—in both Washington and Jerusalem—by considering the history that undergirds today's conundrums, offers some echoes, and makes up at least some of the intellectual infrastructure that may inform decisions in both governments. This study focuses on four key cases in which Israel, after complicated decisionmaking processes, ultimately chose to strike first: the Suez War of 1956, in which the Eisenhower administration was shocked and enraged by Israel's secret collusion with Britain and France to try to topple Egyptian President Gamal Abdul Nasser; the Six-Day War of 1967, in which the Johnson administration contended with repeated Israeli entreaties for a green light to use force after Nasser took a series of increasingly aggressive steps; the 1981 Israeli strike on Iraqi President Saddam Hussein's Osiraq nuclear reactor, in which the Reagan administration was surprised by an Israeli strike; and the 2007 Israeli bombing of a mysterious Syrian nuclear facility known as al-Kibar, in which the George W. Bush administration found itself split over Israel's top-level requests that the United States bomb the Syrian reactor. These cases also track the widening and deepening special relationship between the planet's greatest superpower and an embattled democracy in a volatile and hostile region, from the frigid disdain of President Eisenhower and Secretary of State John Foster Dulles to the far friendlier worldview of their successors in the early 21st century.

This study's first two cases—both full-scale regional wars—were crucibles that helped shape today's U.S.-Israel relationship; the more limited strikes in the second two cases, which we now understand in fairly rich detail, suggest some of the calculations that may help shape Israeli decisionmaking in potential future crises in which Israel finds itself grappling with the apparent nuclear ambitions of a profoundly hostile nearby state. There is much to be learned about Israeli decisionmaking from consideration of other cases in which Israel has used preventive or retaliatory force, such as the 2006 war against Hizballah or the recent Israeli campaigns in Gaza, but this study—of necessity—has chosen a more narrow scope that focuses on clear cases of preemption or preventive force, including two seminal episodes in shaping Israeli views on whether to inform the United States about its upcoming strikes and on two cases that (however imperfectly) offer some parallels to today's Iran standoff.

This study draws upon open sources, ranging from the rich historiography about the Suez War and the Six-Day War to intrepid recent reporting about the 2007 Israeli strike inside Syria and the sometimes quarreling accounts offered by former Bush administration officials. It does not pretend to predict Israeli behavior going forward; rather, it attempts to use historical narrative to inform the reader's understanding of choices both past and present.

Broadly speaking, this study splits Israeli leaders into two categories: confronters and consulters. David Ben-Gurion and Menachem Begin presented the United States with *faits accomplis* in 1956 and 1981, eliciting fury and serious irritation, respectively; by contrast, Levi Eshkol and Ehud Olmert took pains to try to see if Washington would resolve Israel's security dilemmas in 1967 and 2007, which did not produce much by way of forceful American action but yielded considerable dividends of American understanding when Israel chose to take matters into its own hands.

While the historical cases come from quite different times and contexts, several common factors can be seen at play throughout—and policymakers in both Washington and Jerusalem should be mindful of them as they weigh any future decisions about using force against Iran's nuclear program. These factors include the Israeli and U.S. perceptions of the imminence of the threat that may be acted against; the importance of the U.S. interests implicated by potential Israeli military action; the nature of U.S. involvement in the broader Middle East in the time period of potential Israeli military action, including the deployment of U.S. forces in the region; the overall state of play in U.S.-Israel relations, including other sources of bilateral tension or cooperation and recent episodes of strain or partnership; the style and ideology of both the Israeli prime minister and the U.S. president; the nature of U.S. diplomatic efforts to resolve the sources of tension that have given rise to potential Israeli military action; the lessons drawn by both Israeli and U.S. decisionmakers from past episodes involving Israeli preemptive or preventive strikes; and the U.S. administration's overall view of Israel's relative strategic utility in the Middle East.

This study's principal value to policymakers is likely to be in the details of how the episodes of 1956, 1967, 1981, and 2007 unfolded. Those specifics, recounted here with a particular eye to the question of Israeli notification of Washington that is often not directly found in the rest of the literature, may be helpful to policymakers considering Iran-related questions today. A conclusion after the case studies offers some closing reflections.

All observations and interpretations are strictly those of the author.

Washington, DC
October 2013

Case I: The Suez War, 1956

Suez stands apart in the history of U.S.-Israel relations—a moment when Israel's leadership, at the most senior levels, chose both to use force in a truly dramatic fashion, with major implications for U.S. strategy, and to hide the upcoming strike from the United States. So stark and deliberate was the Israeli decision to keep Washington in the dark that on October 28, 1956, the legendary Israeli diplomat Abba Eban had one of the most awkward encounters of his storied career. Eban was then doing double duty as Israel's ambassador to both Washington and the United Nations (UN). That morning, Eban and his deputy found themselves at the State Department, assuring Assistant Secretary of State for Near Eastern Affairs William Rountree of Israel's "defensive posture." The meeting was interrupted by Donald Bergus, the head of the Bureau of Near Eastern Affairs's (NEA's) Israeli-Palestinian affairs office, who entered to pass a note to his boss. Rountree read it aloud: there had been "a massive eruption of Israeli forces around the Egyptian boundary and a parachute drop deep into Sinai." With undiplomatic sarcasm, Rountree told Eban, "I expect you'll want to get back to your embassy to find out what is happening in your country."[1]

All this makes Suez a dramatic outlier—and, in a sense, the starting point for considering the future evolution of U.S.-Israel relations on the question of preemptive and preventive military action.

Throughout the Suez affair, Israel's focus was less on informing the United States—to which Israel saw no upside—and more on work-

[1] Abba Eban, *Personal Witness: Israel Through My Eyes*, New York: Putnam, 1992, p. 258.

ing painstakingly to wrest assurances from its unreliable allies in Paris and London that Israel would not be left alone on the battlefield. In other words, Ben-Gurion's principal anxieties seem to have revolved not around U.S. action but around British and French inaction.

The Suez experience does seem to have left Israeli leaders scalded. It is hard to disentangle precisely how much of the Eisenhower administration's ire at Israel came from U.S. surprise and how much from the simple fact of the invasion, but Israeli leaders ultimately learned that there were real risks and costs to keeping Washington in the dark.

Part of the harsh U.S. reaction was colored by the Eisenhower administration's own perceptions of Israel. The U.S. response—clear condemnation of Israel as an aggressor, threats of sanctions and cutoffs of aid, a determined and ongoing campaign for UN action in both the Security Council and the General Assembly to reverse Israel's moves, and private internal fulminations that occasionally skirted uncomfortably close to the line between exhausted, exasperated pique and outright anti-Semitism—was the product of an administration with its own pronounced view of the Cold War and of Israel's relative utility in advancing U.S. strategy.[2] No other administration in U.S. history has so clearly and consistently regarded Israel with suspicion and driven it so determinedly to give up territory it acquired during wartime.

The basic outlines of the Suez war look no less strange several decades later. The crisis grew out of the rise of Egypt's fiery, pan-Arab nationalist president, Gamal Abdul Nasser, and the anxieties his ascent spurred in Paris, London, and Jerusalem. Nasser, who came to power after the 1952 coup that toppled King Faruq, hoped to mold the Arabs into a major, nonaligned power in one united state. He wanted to break free of Europe's economic grip and destroy the encircling Baghdad Pact alliance forged by the Eisenhower administration, which Nasser saw as an expression of Western hegemony. In 1955, even as the last of Brit-

---

[2] During the crisis, Eisenhower told a friend that he "gave strict orders to the State Department that they should inform Israel that we would handle our affairs exactly as though we didn't have a Jew in America." A few weeks later, Dulles complained about "how almost impossible it is in this country to carry out [a] foreign policy not approved by the Jews." See David Schoenbaum, *The United States and the State of Israel*, New York: Oxford University Press, 1993, pp. 115–119.

ain's old colonial troop presence in Egypt was being removed, Nasser upset the regional balance of power—never terribly stable—by striking an arms deal with Czechoslovakia, raising the prospect that the most populous Arab state might be sliding into the Soviet orbit. His distaste for the West only grew when the United States and Britain wound up clumsily rescinding their offer to help fund Nasser's signature development project, the Aswan High Dam. On July 26, 1956, Nasser struck back—spurring a major crisis by nationalizing the Suez Canal, which was jointly owned by the region's major erstwhile imperial powers, Britain and France.

Perhaps the most belligerent outside power was France, which was falling deeper into a bitter struggle to retain its prized colonial possession, Algeria, and blamed Nasser for fomenting the nationalist rebellion there. If Nasser was toppled, some French leaders reasoned, they would find it easier to hold onto Algeria.

Meanwhile, under the stewardship of a rising young star in Israeli defense and political circles named Shimon Peres, Israel was forging an arms relationship of its own with France, which would ultimately provide the Jewish state with both its nuclear reactor at Dimona and the air force that would let it win the Six-Day War of 1967. U.S.-Israel relations during this period were comparatively frosty; Israel was widely seen at the State Department as a liability impeding Washington's ability to lure the Arab states into the Western camp in the Cold War. France was Israel's main arms supplier, which left Israeli leaders reluctant to alienate their French partners and more prone to follow their lead. Peres's boss, Prime Minister David Ben-Gurion, was also deeply worried about ongoing *fedayeen* (guerrilla) raids into Israel, Nasser's blockade of the strategically crucial Straits of Tiran, and the effects on the region's military balance from Nasser's deepening arms relationship with the Soviet bloc. Ben-Gurion was increasingly anxious to cut Nasser down to size and reassert Israeli deterrence by demonstrating the muscle of the Israel Defense Forces (IDF).

If France and Israel's motives have long been fairly clear, the drivers of British behavior continue to intrigue, divide, and puzzle scholars. What is clear is that Anthony Eden, the British prime minister, disastrously misread Nasser by seeing him as a strutting, blustering "Hitler on the Nile" rather than a genuine nationalist leader. What is also clear is that Eden joined French Prime Minister Guy Mollet and Ben-Gurion in colluding to topple Nasser. In October, Mollet, Eden, and Ben-Gurion met secretly in Sèvres, France, and came to the agreement behind perhaps the modern Middle East's oddest war: Israel would attack Egypt, and Britain and France would then follow suit with their own intervention, which they would call an attempt to separate the combatants. In the end, all three agreed, Nasser should be toppled, and free navigation through the canal should be restored.

Eisenhower and his secretary of state, John Foster Dulles, were deeply troubled by Nasser's actions by the fall of 1956; they had considerable sympathy with Mollet, Eden, and Ben-Gurion's diagnosis (though not their ultimate remedy). Eisenhower and Dulles sent mixed signals to their impatient British and French allies, variously likening Nasser to Hitler and warning that London and Paris "had not yet made their case."[3] But Eisenhower and Dulles did not agree that regime change and gunboat diplomacy were the way to handle the crisis—and their objections became all the more vociferous because their surprise at their friends' behavior was close to total.

Close but not entirely total: As Spiegel notes, Washington did have some hints, and they left Eisenhower and Dulles worried indeed. As the Sèvres collusion reached its climax, Eisenhower and Dulles "had only tidbits of information: tension on the Jordanian-Israeli frontier, continuing Israeli mobilization, and Franco-British buildup in the Mediterranean, a suspicious termination of regular high-level communications with Washington by Paris and London, a sizeable growth in Israeli-French diplomatic radio traffic, a large increase in French Mystère pursuit planes for Israel beyond the number reported to Washington."[4] That was hardly enough to finger Britain and France, but it was enough to spur Eisenhower to warn Ben-Gurion against the use of force.

---

[3] Steven Spiegel, *The Other Arab-Israeli Conflict: Making America's Middle East Policy from Truman to Reagan*, Chicago: University of Chicago Press, 1986, p. 72.

[4] Spiegel, 1986, p. 74.

Because of Sèvres, the Israelis had their own great-power patrons in 1956 and chose not to also ask Washington's blessing—which, of course, would have probably risked the entire campaign.

On October 26, Dulles told UN Ambassador Henry Cabot Lodge that he was worried about pending conflict. As the IDF began to call up its reserves, U-2 surveillance further deepened Washington's worries. (The crisis's leading historian, Keith Kyle, also wryly notes that at least some Americans concluded that the IDF was in the midst of a truly comprehensive mobilization because "their military attaché's driver, who was three fingers short, had been called up.")[5] The U.S. Army attaché in Tel Aviv concluded that the IDF "call-up was on a larger scale than had been ordered since 1948–9" and Israel's War of Independence.[6] He noted that lights were burning in the Israeli Defense Ministry on the eve of the Jewish Sabbath, and his suspicions were heightened by sources who warned that Israel was colluding with the French to target the Straits of Tiran.

If Washington was increasingly convinced that something was up, it was still in the dark about what. When Eisenhower, Dulles, and a few aides reviewed the situation in a Saturday morning meeting on October 27, they concluded that the "most probable direction for an Israeli attack" was the Jordanian-held West Bank, not Suez.[7] At noon Washington time, Eisenhower sent a special demarche to Ben-Gurion—but the Israeli leader put him off, claiming that Israel's regularly scheduled Sunday cabinet meeting had gobbled up the entire day. Ben-Gurion finally received the U.S. ambassador in Tel Aviv at 8 p.m. local time on Sunday, October 28. Moshe Dayan reported that the prime minister was deeply worried about the U.S. message. Whatever his state of mind, Ben-Gurion was far less than truthful with the U.S. envoy, whom he told that Israel had mobilized only "a few units" strictly as a "defensive precaution."[8] That evening, the Central Intel-

---

5 Keith Kyle, *Suez*, New York: St. Martin's, 1991, p. 338.

6 Kyle, 1991, p. 338.

7 Kyle, 1991, p. 338.

8 Kyle, 1991, p. 344.

ligence Agency (CIA) warned that an "attack will be launched against Egypt in the very near future."[9]

The data points were now starting to pile up in Washington: "the exceptionally heavy cable traffic between Paris and Israel," efforts from the usually collegial British Embassy that U.S. policymakers worried were aimed at "keeping us completely in the dark," and a troubling rupture in the usually close contacts in the field between U.S., British, and French military attachés. By this point, Washington felt "almost certain" that Israel's increasingly obvious mobilization meant that it was colluding militarily with France; U.S. policymakers were far less sure about British involvement—and, Kyle notes, eager for it not to be true.[10]

When the French ambassador to Washington and the British chargé d'affaires met with Dulles that Sunday night, they unconvincingly assured the secretary of state that they knew nothing about the mounting evidence of Israeli mobilization—"a form of ignorance," Kyle notes, "that Dulles told the President was almost a sign of a guilty conscience." Dulles warned both envoys not to assume that the administration could not move harshly against Israel even though it was mere days before Eisenhower's reelection bid against former Illinois governor Adlai Stevenson, and the White House made clear that it understood something was afoot by putting out a statement calling Israel's mobilization "almost complete."[11] Dulles ordered his ambassador in London to ask British Foreign Secretary Selwyn Lloyd why Israel was mobilizing—and Lloyd, "with a perfectly straight face, confessed to being in the dark about this."[12]

Israel kept up its own public deception. The marathon cabinet meeting on Sunday ended by releasing a statement explaining that Israel had called up a few reserve battalions due to a sudden confluence of alarming factors: Palestinian guerrilla raids, "the new military alli-

---

[9] Kyle, 1991, p. 345.

[10] Kyle, 1991, p. 345.

[11] Kyle, 1991, p. 345.

[12] Kyle, 1991, p. 346.

ance between Egypt, Jordan and Syria," and Iraq's own mobilization of troops along its frontier with Jordan.[13]

Behind the scenes, Israel tried to walk a somewhat finer line. Peres's biographer reports that, at the last minute, Eisenhower sent "two urgent messages to Ben-Gurion, warning Israel against a military operation"—although the president still thought Israel had Jordan in its gunsights, not Egypt. Ben-Gurion's reply was "carefully worded, so as not to lead [Eisenhower] astray," but he declined to "promise that Israel would refrain from military action."[14]

In stark contrast to later episodes of Israeli preemptive or preventive strikes, the day before the outbreak of combat in 1956, Israel's ambassador to the United States was not pounding on doors at the White House and the State Department but perambulating the links of a nearby golf course. Eban did not just seem to be in the dark; he *was* in the dark, despite having just come back from Israel, where a conscious decision must have been made not to mention the imminent attack to the Jewish state's most important ambassador. When Dulles summoned Eban to the State Department, the Israeli diplomat "found the secretary staring at a huge map of the Israeli-Jordanian frontier."[15] The Middle East would be at war in mere hours, and—thanks to deliberate secrecy from two North Atlantic Treaty Organization (NATO) allies and one aspiring U.S. client—the secretary of state was not even looking at a map of the right country.

But on Monday, October 29, the subterfuge ended, and the war began. That evening, Israel summoned the British and French military attachés to its defense ministry and told them that Israeli paratroopers had entered Egypt. Israel's first official communication of its invasion, in other words, went to its Sèvres partners, not to the United States. "Nasser was taken completely by surprise," as Kyle notes, and Eisenhower was not far behind.[16]

---

[13] Kyle, 1991, p. 347.

[14] Michael Bar-Zohar, *Shimon Peres: The Biography*, New York: Random House, 2007, p. 156.

[15] Spiegel, 1986, p. 74.

[16] Kyle, 1991, p. 350.

Viewed today, the Eisenhower administration's response to Suez is startlingly sharp—and at least part of the ferocity of the U.S. response was due to Eisenhower and Dulles's fury at being surprised. (Eisenhower reportedly seethed that his erstwhile allies could boil in their own oil.) Washington took Suez directly to the UN Security Council, pushing a U.S.-drafted resolution that called for "Israeli withdrawal, no force by other states, and no aid to Israel."[17] The draft text was killed off on the second day of the war by twin vetoes by Britain and France, both permanent members of the council. Fuming, the administration pivoted instead to the General Assembly, where Dulles warned that "the violent armed attack by three of our members upon a fourth" was "a grave error" that flouted the principles of the UN Charter.[18]

Eisenhower's rage only rose as Soviet tanks rolled into the streets of Budapest to put down the 1956 Hungarian uprising, leaving him in the unhappy position of condemning Soviet aggression in Hungary while managing the tripartite invasion of Egypt by his friends. Election day in the United States and the last round against Democratic presidential nominee Adlai Stevenson seemed almost an afterthought. The day after his reelection, Eisenhower warned Israel to get out of the Sinai or face "UN condemnations, attack by Soviet 'volunteers,' [and] termination of all U.S. governmental and private aid."[19] The next day, Ben-Gurion reluctantly agreed that the IDF would pull out. Israel slowed that process into February 1957, when Eisenhower delivered a televised speech that warned Israel's leadership that its time was up. As Spiegel notes, "the United States had brought Israel to withdraw without substantial public commitments in return."[20]

Israel had entered the Suez crisis reliant on France for its arsenal, and by joining up with France and Britain, it shifted its focus away from the great power whose support it wanted and toward the great powers whose support it had. But Ben-Gurion may not have fully real-

---

[17] Spiegel, 1986, p. 75.

[18] Spiegel, 1986, p. 76.

[19] Spiegel, 1986, p. 77.

[20] Spiegel, 1986, p. 81.

ized the depth to which top-tier U.S. national security interests were engaged by the Suez episode—and the stakes were only compounded by the alarming Soviet thrust into Hungary. Nor did Ben-Gurion have a deep reservoir of emotional support for Israel to draw upon from Eisenhower or the frosty Dulles.

The Suez crisis taught the Israeli national security establishment a lasting lesson: Surprising the United States has serious consequences. "It was clear we were at physical war with Egypt, and almost at emotional war with the United States," Eban remembered. "All contact between the [Israeli] embassy and the State Department was sundered." When one of Eban's deputies reached out to his Middle East counterpart at the State Department, the Israeli diplomat was coldly told that the only matter their two governments had to discuss was the evacuation of U.S. citizens from Israel.[21]

The memory of that "emotional war" stayed with Ben-Gurion and the rest of the senior Israeli leadership for decades to come. Preventive war that startled the United States and interfered with its vital interests carried a heavy price, including the largely unrewarded return of the territorial gains that Israel made from the IDF's tactically skillful 1956 campaign. Surprising the United States may have been the price that Ben-Gurion chose to pay for working with Britain and France to reduce the threat from Nasser in 1956. But future Israeli policymakers would conclude that their great-power patron of choice lay not across the Mediterranean, but across the Atlantic.

---

[21] Eban, 1992, p. 258.

Case II: The Six-Day War, 1967

If in 1956, Israel went nowhere in Washington for fear of being told no, in 1967 Israel went almost everywhere in Washington in hopes of being told yes.

The backdrop to the Six-Day War was months of mounting tension—including *fedayeen* attacks on Israel, a dramatic dogfight over Damascus between Israeli and Syrian planes, and a major Israeli raid on the Jordanian town of Samu. On May 14, 1967, the crisis began in earnest as Nasser sent Egyptian troops back into Sinai. On May 16, he took a much larger escalatory step by asking the United Nations Emergency Force (UNEF), the UN peacekeeping force installed after the Suez war, for a partial withdrawal, leaving the region near the Egyptian-Israeli frontier while leaving the blue helmets in the Gaza Strip and Sharm al-Shaykh.

The next day, Israeli Prime Minister Levi Eshkol—a supple, humorous, and shrewd figure who lacked Ben-Gurion's bombast and obstinacy but also his charisma and towering credibility on national security issues—received a letter from Lyndon Johnson redolent of the aftermath of Suez. During the decade of relative quiet that followed Suez, Israel had come to rely on Washington's 1957 commitments—echoed by Dulles and others as part of Israel's withdrawal from Sinai—to keep the strategically crucial Straits of Tiran open. But now, LBJ pointedly noted, "I cannot accept any responsibilities on behalf of the

United States for situations which arise as the result of actions on which we are not consulted."[1]

From the outset of the 1967 crisis, an unmistakable message was sent at the very highest levels: Washington did not intend to be caught as flatfooted as it was 11 years ago. Johnson was distinctly firm because he was decidedly preoccupied. As his senior advisers debated further U.S. involvement in Vietnam, he had scant U.S. military assets to spare if Israel were to get itself into mortal trouble and require U.S. military intervention to avoid being overrun. The last thing the Defense Department wanted was another U.S. military intervention. The U.S. Intelligence Community was always fairly confident in Israel's ability to beat back even a combined Arab attack; indeed, America's spies were a lot more confident about Israel's might than were Israel's citizens. But however Israel handled the 1967 crisis, the United States did not want to be surprised. Israeli decisionmaking was powerfully shaped by the ongoing U.S. attempts to resolve the conflict short of war—even as Israel endured one of the most harrowing crises in its history.

But if the Johnson administration did not want to be surprised, it was also distracted enough to find itself faced with some dicey alternatives. Matters were made worse early on by a major blunder by UN Secretary General U Thant, who told Nasser that he would not agree to a partial withdrawal of UNEF—thereby pressing the Egyptian leader to either tolerate the uninterrupted continuation of UNEF's mandate or to call for it to be thrown out wholesale. Unwilling to risk howls of opprobrium from his Arab rivals, Nasser chose the latter, and suddenly the buffer zone between the 1956 belligerents was no more. Johnson was "shocked" and "puzzled" by Thant's retreat.[2] On May 22, Nasser crossed a clear Israeli red line by announcing that he was closing the strategically vital Straits of Tiran to Israeli shipping—something Israel had long made clear was *casus belli*.

For our purposes, what is striking is that in neither case did the United States seem to have made exceptional exertions to head off

---

[1] William B. Quandt, *Decade of Decisions: American Policy Toward the Arab-Israeli Conflict, 1967–1976*, Berkeley, Calif.: University of California Press, 1977, pp. 39–40.

[2] Spiegel, 1986, p. 137.

fairly predictable and major trouble at the pass. Washington seems not to have pressed Thant to keep UNEF (or as much of it as possible) deployed as a brake on tensions. Moreover, as William Quandt notes, we have found "no sign of any direct American approach" to ask Nasser to keep the Straits of Tiran open until the day he declared them closed—and even then, the U.S. contact consisted of a letter from Johnson filled with blandishments about U.S. friendship with Egypt while warning Nasser off of actions that might trigger war.[3] But Washington was fairly direct in its initial dealings with Israel over the prospect of preemption. On May 17, Undersecretary of State Eugene Rostow for the first time sounded what would become a familiar theme in a meeting with Israel's ambassador to Washington, Avraham Harman: Israel "will not stand alone," the U.S. diplomat averred, unless it took unilateral military action.[4]

The few, somewhat scattered U.S. actions up to this point had clearly failed to slow the rush to escalate. Indeed, the Johnson administration's seeming confusion about the commitments that Eisenhower's team had made to Israel after Suez only ratcheted up Israeli anxieties even higher. As Quandt notes, U.S. diplomacy "went into high gear" only after Washington's earlier, piecemeal attempts to cool the temperature had failed—and once the prospect of Israeli preemption was "acutely real." When the Israeli cabinet met on the morning of May 23 to consider Nasser's blockade of the Straits of Tiran, Abba Eban—now Israel's foreign minister, and quite probably with the echoes of Eisenhower and Dulles's Suez rebukes still ringing in his ears more than a decade later—urged his fellow ministers for a few days to try to rally international support. Eban read to his colleagues a cable from the Israeli Embassy in Washington in which Johnson formally asked for a 48-hour delay.[5] Walworth Barbour, the U.S. ambassador to Tel Aviv, reiterated the U.S. request. Eban warned his fellow cabinet members

---

[3] Quandt, 1977, p. 41.

[4] Michael Oren, *Six Days of War: June 1967 and the Making of the Modern Middle East*, New York: Oxford University Press, 2002, p. 77.

[5] Michael Brecher, *Decisions in Israel's Foreign Policy*, New Haven: Yale University Press, 1975, pp. 378–379.

against a sequel to Suez. Moshe Dayan, one of the cabinet's most formidable military figures, grumbled about the time-wasting prospect of "banging on the doors" of the United States and other great powers; IDF Chief of Staff Yitzhak Rabin worried that Egypt was digging into its positions in the Sinai desert.[6] But the military men's qualms were overruled, and Eban was dispatched to Washington for urgent consultations.

During this phase, the Johnson administration was working to hold Israel back—affirmatively trying to stop Israel from striking preemptively. To stay Israel's hand, Washington started working on an international effort to reopen the Straits of Tiran, a push in the UN Security Council to get the Straits open again, or even to consider what came to be known as the "Red Sea Regatta"—a multinational flotilla of ships that would pass through the straits and reaffirm their importance as a freely navigable waterway. The administration devoted considerable energy to multilateral action (in the UN and outside it) precisely because it had, "virtually from the beginning," ruled out unilateral Israeli action.[7] But the Pentagon began slowly chipping away at the objections to Israeli preemption—to some degree because it blanched at the prospect of having to oversee the Red Sea Regatta and to some degree because U.S. military planners became increasingly convinced that an Israeli campaign could end the crisis effectively. (Johnson, on the other hand, remained deeply worried that Israel would find itself in over its head and insisted that the Pentagon and Intelligence Community reexamine their rosy assessments—which, of course, turned out to be on the mark.)[8]

Eban landed in Washington on May 25, after stopping first in Paris (where Israel's old Suez ally further alarmed Israeli leaders by warning them not to strike first) and London (which remained mum). Eban was greeted at New York's Kennedy Airport by Israel's ambassador to the United States, Avraham Harman, who was clutching a new

---

[6] Oren, 2002, pp. 89–90.

[7] Quandt, 1977, p. 45.

[8] Richard B. Parker, *The Politics of Miscalculation in the Middle East*, Bloomington: Indiana University Press, 1993, p. 115.

cable from Jerusalem warning that an Egyptian attack was imminent.[9] A dismayed Eban was given new instructions: to ask the Americans to declare that an Egyptian attack on Israel "would be viewed as an attack on the United States."[10] The foreign minister, who fretted that IDF Chief of Staff Rabin and others were exaggerating the Egyptian threat, visited the Pentagon and the State Department the next day, where he was met with skepticism that Nasser was about to strike and assurances that Israel need not rush to war. Defense Secretary Robert McNamara, Director of Central Intelligence Richard Helms, and others all told Eban that Israel "would easily win" in less than a week if fighting broke out.[11] Secretary of State Dean Rusk warned Eban that "a planned Israeli preemptive strike" would be "a horrendous error."[12]

Later on May 26, another Israeli diplomat, Ephraim Evron, made his way to the White House to urge LBJ to find time to see Eban. Evron was called into the Oval Office for an informal chat. Johnson told him that "he did not believe" that Israel would strike unilaterally, which would unleash consequences for which Washington would not be responsible.[13] "Israel is not a satellite of the United States," Johnson said, "nor is the United States a satellite of Israel."[14]

At 7 p.m. on May 26, Eban's meeting with Johnson began, with McNamara, Rusk, and other senior officials sitting in. LBJ asked for time to work with Congress and U.S. allies on a plan to reopen the Straits. Johnson thrice repeated the mantra coined in the principals' meeting earlier that day: *"Israel will not be alone unless it decides to go it alone."*[15] Echoing his earlier signal to Evron, he added that he "could not imagine Israel making a precipitate decision."[16] The president also

---

9 Spiegel, 1986, p. 138.

10 Quandt, 1977, p. 48.

11 Quandt, 1977, p. 50.

12 Oren, 2002, p. 108.

13 Quandt, 1977, p. 53.

14 Oren, 2002, p. 114.

15 Oren, 2002, p. 115; emphasis in original.

16 Quandt, 1977, p. 53.

assured Eban that he would make every effort to open the Straits of Tiran to Israeli shipping. In an *aide memoire* for Eban to take back to Jerusalem, LBJ emphasized "the necessity for Israel not to make itself responsible for the initiation of hostilities. Israel will not be alone unless it decides to go alone. We cannot imagine that it will make this decision." But Johnson could indeed imagine exactly that. After enduring a presidential handshake of crushing power, Eban left, and Johnson was left alone with his most senior advisers. "I've failed," the president said. "They'll go."[17]

To try to stave that off, LBJ played for time. He sent a letter underscoring his warning to Eban: "Israel just must not take preemptive military action and thereby make itself responsible for the initiation of hostilities."[18] Eban was staggered by his Oval Office meeting, which he thought underscored LBJ's impotence, paralysis, and defeatism.[19] Johnson asked Eshkol to commit not to act for two more weeks—an excruciating interval for an Israeli citizenry already worried about a second Holocaust. On May 28, a reluctant Eshkol promised to hold off for "a week or two."[20]

Meanwhile, increasingly anxious Israeli policymakers wondered whether the United States actually wanted Israel to preempt and wrap up the crisis quickly. Johnson's tortured "go it alone" mantra was, after all, a far cry from Eisenhower's blunt condemnations of Israeli aggression. On their flight home, Eban and an aide reportedly tried to puzzle out the phrase's meaning and concluded that "it meant that if Israel struck first it would be on its own, but that did *not* mean the United States would *oppose* it."[21] As Quandt wryly notes, "It did not require a Talmudist to read into the phrase the hint of a 'green light' to Israel—

[17] Quandt, 1977, p. 54.

[18] Spiegel, 1986, p. 143.

[19] Oren, 2002, p. 115.

[20] Quandt, 1977, p. 56.

[21] Parker, 1993, p. 114; italics mine.

and there were plenty of Talmudists in Israel who were convinced that United States policy was precisely that."[22]

Eban warned his cabinet colleagues not to overread Johnson's comments, which the foreign minister took as a flat "no" to preemption. A bitterly split cabinet agreed to give Washington up to three weeks to work.[23] To try to figure out whether the U.S. light was green, amber, or red, Eshkol personally dispatched Mossad chief Meir Amit back to Washington to seek further clarity.

Earlier in the crisis, a senior U.S. official had flatly warned Amit: "If you fire the first shot, you're on your own."[24] Now, with nerves frayed and war looming, Amit met with Pentagon officials. They continued to sour on Johnson's multinational flotilla, which was finding scant takers among U.S. allies and in a Congress still perturbed by the Gulf of Tonkin resolution that gave Johnson a free hand on Vietnam. Amit told Defense Secretary McNamara that the war would take Israel just two days to win and noted that he would return to Jerusalem and recommend that Israel strike. "I read you loud and clear," McNamara said calmly. "This was very helpful."[25] Amit reported back to Jerusalem that the Red Sea Regatta "was unlikely ever to sail," which implied that diplomacy had run its course and Washington might well "accept Israel's taking matters into its own hands."[26] On May 31, a reporter asked Rusk whether Washington was trying to restrain Israel, and the secretary of state replied, "I don't think it is our business to restrain anyone"—a morsel savored by Israel's edgy leadership.[27]

Meanwhile, an already grave crisis worsened: Jordan's moderate King Hussein flew to Cairo to sign a mutual-defense deal with Nasser; Nasser continued to get spectacularly bad military advice from his gen-

---

[22] Quandt, 1977, p. 56.

[23] Oren, 2002, p. 124.

[24] Oren, 2002, p. 146.

[25] Oren, 2002, p. 147. McNamara later said, "we were absolutely opposed to preemption."

[26] Spiegel, 1986, p. 146.

[27] Quandt, 1977, p. 57.

erals, especially army chief Marshal Abdul Hakim Amr;[28] the Soviets grumbled about imminent Israeli strikes; Israel found itself stretched to maintain the IDF's ongoing mobilization of reservists; Eshkol brought his rightist opposition into a new, wall-to-wall national-unity government featuring Moshe Dayan, a wartime *consigliere* if ever there was one, as defense minister; and Nasser made his most belligerent statement yet, declaring that "Israel's existence in itself is an aggression."[29]

On June 2, Evron met with Walt Rostow of the White House's National Security Council (NSC) staff "to make sure that Johnson understood that time was very short and that Israel might have to go to war."[30] Rostow asked how much time they had; Evron alluded to the old two-week promise from Eshkol, which would mean June 11. On June 2, Israel's ambassador to Washington, Avraham Harman, met with Rusk, who again warned the Israelis not to strike first. Even Rusk admitted in a cable to regional ambassadors that unless Washington could reopen the Straits of Tiran, it could no longer restrain Israel.[31]

By now, many at the White House had had enough time to resign themselves to the likelihood of war. Johnson himself was starting to conclude that the Red Sea Regatta was unworkable. On the NSC staff, Harold Saunders warned that "holding Israel back" would make the United States responsible for its long-term security. "The only other choice is to let the Israelis do this job themselves," he argued. "We ought to consider admitting that we have failed and allow fighting to ensue."[32]

Even Eban had by now concluded that the regatta was dead and that war was inevitable. Dayan was egging the government toward war. It now all came down to Eshkol. Throughout the crisis, Israel's mild-mannered, underrated prime minister tried to give Washington all the time he could. "We will still need Johnson's help and support," Eshkol

---

[28] See Parker, 1993, pp. 59–97.

[29] Spiegel, 1986, p. 148.

[30] Quandt, 1977, p. 58.

[31] Spiegel, 1986, p. 148.

[32] Oren, 2002, p. 165.

told his generals, who were straining at the leash. "I want to make it clear to the president, beyond a shadow of a doubt, that we have not misled him; that we've given the necessary time for any political action designed to prevent the war. Two days more or less won't sway the outcome!"[33]

Late on Saturday night, June 3, Eshkol gathered his high command at his home in Jerusalem to hear from Harman and Amit, who had flown back from Washington together. Both envoys reported that "there was no chance of unilateral United States action nor of successful multilateral action. The conclusion was inescapable: Israel was on its own."[34] Eshkol finally concluded that time was up. "I'm convinced," he told what was now a war cabinet, "that today, we must give the order to the IDF to choose the time and the manner to act."

Israel had finally decided on preemptive war. "Washington was not informed of the decision," Quandt notes—but it could hardly have been surprised.[35] Rabin and Dayan chose to strike starting at 7 a.m. on Monday, June 5. That morning, Israeli planes destroyed Nasser's air force on the ground, and the Six-Day War began.

In large part because of the extended dance between Washington and Jerusalem, LBJ's response to Israel's strike was almost 180 degrees different from Eisenhower's wrath 11 years before. In his memoirs, Johnson admitted to "regret that Israel decided to move when it did" but flatly rejected "the oversimplified charge" that Israel had acted aggressively.[36] Johnson had discouraged preemption and feared its consequences, but once he discovered that the Pentagon had been quite right in its healthy respect for the IDF's prowess, he let himself breathe a sigh of relief that the United States would not have to pull Israel's chestnuts out of the fire. If Eisenhower had been worried at a moment of great surprise that *preventive* Israeli military action would make the United States lose Cold War ground by looking like a hypocritical cod-

---

33 Oren, 2002, p. 152.

34 Quandt, 1977, p. 59.

35 Quandt, 1977, p. 59.

36 Lyndon B. Johnson, *The Vantage Point: Perspectives of the Presidency, 1963–69*, New York: Holt, Rinehart, & Winston, 1971, p. 297.

dler of an aggressive client, LBJ had been worried that *preemptive* Israeli military action after a long buildup could force the United States into another intervention to save a friend that had lost a gamble. But now that Eshkol's gamble seemed to be paying off, as the Pentagon and the CIA had predicted it would, LBJ did what Eisenhower had refused to: gave Israel time and space to work.

The Johnson administration backed a cease-fire and concentrated on keeping the Soviets from intervening to help out their own client, who was faring rather less well than Washington's. A frantic Nasser did his cause little good by claiming that the United States was fighting alongside the Israelis. In the event, the window for wartime crisis management slammed shut with stunning speed. With something somewhat shy of America's blessing, Israel had moved unilaterally in six sharp days of combat to reshape the Middle East for decades to come.

Interestingly, before Israel struck, at least some U.S. policymakers had tried to use the prospect of Israeli preemption to warn Egypt off its reckless course. As the former U.S. diplomat Richard Parker notes, he and his colleagues "made no attempt to hide their belief that Israel would strike" if Cairo hewed to its hard line. "That was made clear to Egyptian officials at various levels in Cairo," Parker notes, "but they seemed not to be worried by that eventuality."[37]

If the diplomatic trauma of 1956 predisposed Israeli policymakers never to strike first without U.S. approval, the military victory of 1967 highlighted for Israeli policymakers the frustrations of seeking approval from a balky superpower. Israel had been caught up in interagency tensions between a State Department keen to reach for multilateral solutions and a Defense Department leery of anything that risked additional U.S. deployments as the nation sank deeper into Vietnam. Johnson had proven difficult to read, reaffirming his support for Israel's right to free navigation in the Straits of Tiran even as he warned against unilateral action and demanded more time to let diplomacy run its course. He had also come up with a catchphrase ("Israel will not be alone unless it decides to go alone") that still stands as a compact masterpiece of confusing diplomatic prose. But in the context of

[37] Parker, 1993, p. 112.

Nasser's steadily more ominous moves, Israel rightly read some of this ambiguity to suggest that the Johnson administration was not as mortally opposed to Israeli *preemption* as the Eisenhower administration had been to Israeli *prevention*. The immediately menacing nature of Nasser's actions, including blockading the Straits of Tiran, presented U.S. policymakers with a more compelling and urgent case for Israeli preemptive action than had the longer-term threat that Nasser's rising regional profile had posed to Israel in 1956. The very act of ongoing and close consultation also had the important underlying net effect of leaving the administration satisfied that Israel had not disregarded vital U.S. interests or legitimate U.S. concerns.

The overall result was an ongoing U.S.-Israeli interaction that was certainly more frustrating than the Suez diplomacy (or nondiplomacy) but also far less jarring to the emerging U.S.-Israel special relationship that had begun under President John F. Kennedy. The context was also dramatically different: Ties between the two nations were now significantly closer than they had been in 1956. The United States had largely given up its attempt to pull the nonaligned Nasser into the Western orbit in the Cold War, and Kennedy had, in August 1962, set the precedent that led to today's massive arms relationship by agreeing to sell Israel a state-of-the-art weapon system, the HAWK antiaircraft missile.[38] Israel had far more to lose by antagonizing the United States, and the United States had far more leverage over the Jewish state. But a new dynamic had also sprung up between the two sides, characterized by far more frequent consultations that made May 1967 into a sharp strain on an emerging standard operating procedure rather than a point of departure. If Israel's determination to keep a great power in its corner had helped drive it to join with Britain and France in surprising the United States over Suez, Israel's yearning for superpower patronage and protection helped induce it to pursue a far more consultative course in 1967.

But it is important not to overstate the case. Johnson was not surprised by Israel's airstrikes, but he was not enthusiastic about them

[38] See Warren Bass, *Support Any Friend: Kennedy's Middle East and the Making of the U.S.-Israeli Alliance*, New York: Oxford University Press, 2003.

either. When Eban returned to the Oval Office in October 1967, he found the president still clearly upset. He had thought Israel's decision to strike "unwise" back in June, LBJ said, and the IDF's stunning victory had not changed his mind. Israel's decisions had forced him to confront "the most awesome decisions he had taken since he came into office."[39] It was not an entirely happy conversation. It was not a dressing-down, however, but a conversation among friends. And it took place in the Oval Office, not the Security Council.

---

[39] Schoenbaum, 1993, p. 155.

Case III: The Raid on Osiraq, 1981

On June 7, 1981, eight U.S.-made F-16 jets arced skyward from Israel on one of the most daring missions in the history of the Jewish state. Accompanied by six F-14 fighters for additional cover and several F-15s rigged out for midair refueling, the F-16 squadron raced eastward over Saudi and Jordanian airspace, across the Gulf of Aqaba, and lunged into Iraq. The Israeli pilots' target was a French-built Tammuz 1 nuclear reactor known as Osiraq, located at al-Tuwaitha, some 10 miles southwest of Baghdad. As the sun sank over the Iraqi capital, the F-16s made a first pass over Osiraq, dropping bombs that blew holes in the reactor's dome. Then a second wave of F-16s swooped past, pelting additional munitions inside the holes with what one stunned French onlooker called "stupefying accuracy." With Osiraq in ruins, the Israeli jets streaked homeward.[1] And while the United States knew of Israel's deep concern about Iraq's nuclear program, the raid itself came as a surprise to the United States. Washington was not informed until the Israeli jets were already on their way home.

There was widespread speculation that Israel had a nuclear arsenal dating back to around the time of the Six-Day War.[2] Israel has con-

[1] Howard Sachar, *A History of Israel, Volume II: From the Aftermath of the Yom Kippur War*, New York: Oxford University Press, 1987, p. 127.

[2] Nuclear Threat Initiative, "Country Profile: Israel," August 2013. On Israel's nuclear program, see Avner Cohen, *Israel and the Bomb*, New York: Columbia University Press, 1998. More popular accounts are found in Michael Karpin, *The Bomb in the Basement: How Israel Went Nuclear and What That Means for the World*, New York: Simon & Schuster, 2007, and Seymour Hersh, *The Samson Option: Israel's Nuclear Arsenal and American Foreign Policy*, New York: Random House, 1991.

tinued to hew to the mantra that the young Shimon Peres had coined during the Kennedy administration, averring that it would not "be the first" to introduce nuclear weapons into the Middle East—but nobody in Jerusalem wanted to see Iraq become the second.

Osiraq had cost Iraq an estimated $250 million. The reactor was part of a budding nuclear relationship with France dating back to the mid-1970s, driven by the Valéry Giscard d'Estaing government's interest in access to Iraq's oil supply. In 1971, Iraqi dictator Saddam Hussein ordered a group of nuclear physicists "to start a nuclear energy program as the cover for a weapons program." The program accelerated in 1976, when Iraq signed a deal with France for a reactor intended for military purposes.[3]

It was not just Israel that was concerned at the prospect of a nuclear-armed Saddam; the United States also worried about the idea of letting a radical, Baathist dictatorship closely aligned with the Soviet Union get the bomb. As one U.S. official put it, "Our worries reflected the quality of the regime as much as specific [nuclear] programs."[4] Iraq also had several options for delivering any warheads, including Mirages, MiG-23s, TU-22 bombers, and Scud missiles.

As Iraq moved ahead with its nuclear program, the United States urged France to stand down, as well as pressuring Italy, another NATO ally helping supply Saddam. But U.S. protests to Paris and Rome fell on deaf ears, and senior U.S. officials candidly admitted that diplomatic efforts to stop the Iraqi program were a steep uphill climb. "We really did try diplomatically," noted Samuel Lewis, then the U.S. ambassador to Israel, "but we had no luck with the Italians and the French."[5]

In the wake of the 1973 Arab-Israeli war, the United States was now providing Israel with higher levels of aid. As the U.S.-Israel relationship deepened, the United States was given ample warning about the depths of Israel's alarm over Osiraq. Some ten months before the raid, Prime Minister Menachem Begin called Lewis in for the first

---

[3] Kenneth Pollack, *The Threatening Storm: The Case for Invading Iraq*, New York: Random House, 2002, p. 173.

[4] "Attack—and Fallout," *TIME*, June 22, 1981, Vol. 117, No. 25, p. 28.

[5] Samuel Lewis, interview with the author, Washington, D.C., December 27, 2012.

time to warn of his deep concern about the Iraqi reactor. Over the next three months, Lewis recalls having dozens of high-level conversations about Osiraq, usually with Begin directly. The U.S. Embassy in Tel Aviv sent cable after cable back to Washington about these conversations. U.S. officials described Begin as convinced, notwithstanding the skepticism of outside nuclear scientists, that Iraq would enter (in the latter-day words of Ehud Barak) a sort of "zone of immunity" once the reactor went critical; thereafter, Begin feared, Israel would not be able to destroy Osiraq without unleashing a cloud of fallout that would kill thousands of Iraqi children. Begin urged the United States to stop the project well before what he considered that key tipping point, including coming down hard on the French, the Germans, the Italians, and others.

Lewis and his colleagues, in response, sought to calm Begin's anxieties. "We tried to be very reassuring," Lewis recalls. U.S. intelligence services regularly shared their assessments of Iraq's progress. Some U.S. officials judged that Iraq was some two to three years away from being able to bring the reactor critical. "As the year progressed," Lewis later recalled, "we kept telling Begin, 'There's ample time, don't worry.'"[6] But Begin was hardly mollified; if anything, he became more agitated. Israeli officials insisted that Iraq was making swifter progress than the Americans realized, leaving Saddam Hussein perhaps a year away from the reactor becoming critical. (Even the Israelis admitted it would be some time beyond that before Iraq could have an actual nuclear arsenal.)

Meanwhile, U.S. diplomats noticed that more and more Israeli journalists were starting to write about the Iraqi bomb threat. Lewis and his colleagues considered this "clearly a calculated campaign to convince us publicly" that Osiraq was gravely dangerous. Lewis added, "We were reporting all of this back to Washington—and getting back, 'Don't worry, we're doing everything we can.'"

Lewis decided to highlight the tensions over Osiraq after Ronald Reagan bested Jimmy Carter in the 1980 elections. Knowing that a series of transition teams would be preparing analyses for the incom-

---

[6] Lewis, 2012.

ing administration, Lewis made a series of secure phone calls back to Washington to make sure the new Reagan team was provided with a history of the Osiraq issue—something he hoped would alert the incoming administration that the reactor was a real problem. Lewis was assured that a detailed Osiraq section was put into a key transition document. But as he recalls, the document was subsequently slapped with an extremely high level of classification, leaving it in a compartment into which very few incoming officials had been read. Few senior officials seem to have ever seen it. Secretary of State Alexander Haig later told Lewis he never saw it. "It got so overclassified that it stayed in a compartment and got lost in the shuffle," Lewis sighed.[7]

The depth of Israel's concern may be seen in a series of covert attempts to take matters into its own hands. Western intelligence agencies suspected Israel's Mossad of involvement in a 1979 attempt to blow up the reactor core while it was still being manufactured near Toulon, as well as the 1980 killing of the head of the Iraqi nuclear program in his hotel room during a visit to Paris.

As Osiraq came closer to fruition, Begin convened his Ministerial Defense Committee on Security Affairs to start weighing a full-blown air assault. The committee was initially split. Foreign Minister Yitzhak Shamir, Agriculture Minister Ariel Sharon, and others backed a preventive strike. So did IDF Chief of Staff Rafael Eitan.[8] But other ministers were skeptical, including Deputy Prime Minister Yigael Yadin, Interior Minister Yosef Burg, and Education Minister Zevulun Hammer, who reportedly "felt that [an] attack would damage relations" with the United States.[9] The hawks prevailed, and Begin began a closely held planning process. In November 1979, Begin issued a directive asking the IDF General Staff to start exploring different options for destroying Osiraq.[10] The ultimate decision about when to strike was

[7] Lewis, 2012.

[8] Schoenbaum, 1993, p. 277.

[9] "Attack—and Fallout," 1981.

[10] Chaim Herzog (updated by Shlomo Gazit), *The Arab-Israeli Wars: War and Peace in the Middle East from the 1948 War of Independence to the Present*, New York: Vintage, 2005, p. 342.

his. After a few false starts, including a tussle in May after reading in Labor Party leader Shimon Peres (who shared Yadin, Burg, and Hammer's reservations), the Israeli military began planning what came to be known as Operation Opera. The raid was set for June 7—three weeks before Israeli elections.

The Iraqi nuclear program was particularly awkward for Peres, the leader of the opposition Labor Party. Begin worried that Peres might defeat him in Israel's June 30, 1981, elections and then lack the nerve to attack Osiraq. While Begin moved ahead with plans for the raid and the elections drew near, France elected Francois Mitterrand as its new president. Peres warned Mitterrand, a fellow member of the Socialist International who considered himself a close friend of Israel's and had blasted Giscard d'Estaing's proliferation policies, that an additional shipment of highly enriched uranium would let Iraq enter the nuclear club. "If I am elected president, France will not deliver the second shipment," Mitterrand promised.[11]

Peres took the good news back to Jerusalem, and Mitterrand's aides began drawing up a plan to distance France from Iraq's nuclear program, but Begin was unimpressed. Peres urged Begin not to proceed with the bombing, fearing "an international diplomatic quarantine," but Begin brushed away his concerns.[12] Israel's initial plan called for bombing Osiraq on the day of Mitterrand's inauguration; that, at least, Peres was able to shift. For operational security, Peres stayed away from Mitterrand's day of triumph, even after the bombing date was pushed back.[13]

Peres finally put his concerns about the possible raid in writing to Begin, urging him to refrain. "I speak out of experience," Peres noted—perhaps an allusion not just to his decades of national security experience, but also to the Suez-era perils of surprising superpowers. He was deeply concerned about international isolation. "Israel would be like

---

[11] Michael Bar-Zohar, *Shimon Peres: The Biography*, New York: Random House, 2007, p. 353.

[12] Sachar, 1987, p. 126.

[13] Bar-Zohar, 2007, p. 353.

a lone tree in the desert," Peres warned.[14] He argued that Osiraq was not yet capable of producing weapons-grade uranium or plutonium, that stopping the French shipment of additional enriched uranium would obviate Saddam's ability to attain nuclear arms "in the immediate future," that the right time to strike the reactor was after it was actually "hot," and that further covert action could slow down the Iraqi program. The letter reinforced Begin's belief that Peres would prove a feckless prime minister and hardened Begin's decision to strike. But years later, Peres remained convinced that he had been right: The raid had driven Saddam to bull ahead with his nuclear arms program with greater secrecy and determination. Had Saddam not made the epic strategic blunder of invading Kuwait in August 1990 and arraying a global and regional coalition against him, Peres mused, "he might well have reached his destination" of becoming a nuclear power despite—and in part because of—the 1981 bombing.[15]

Peres paid a considerable political price for his skepticism about the Osiraq raid: Begin leaked his letter, which was promptly pounced on as pusillanimous. (Decades later, Peres was still smarting over the leak.) Boosted in part by the success of the raid, Begin's rightist Likud Party edged out Peres's Labor Party in the elections just weeks after the bombing.

Even as Peres tried to work the Paris track, U.S. diplomatic efforts and intelligence exchanges continued—to scant avail. In February, Yehoshua Saguy, the head of Israeli military intelligence, traveled to Washington to try to alert the CIA to Israel's suspicions and try to close the gap between the relatively sanguine U.S. assessment and the far more alarmed Israeli one. Haig later told Lewis that Saguy never raised Osiraq with him. When Haig first traveled to Israel, he met with Begin and his entire cabinet and conveyed his sympathy with Israel's security concerns—but, as Lewis recalls, neither side raised the subject of Osiraq. Begin and his ministers may well have taken Haig's

---

[14] Bar-Zohar, 2007, p. 354.

[15] Shimon Peres, *Battling for Peace: A Memoir,* New York: Random House, 1995, p. 184. See also Hal Brands and David Palkki, "Saddam, Israel, and the Bomb: Nuclear Alarmism Justified?" *International Security,* Vol. 36, No. 1, Summer 2011, pp. 133–166.

warmth to mean that the new U.S. administration would not be overly unhappy if Israel later lashed out against Iraq.

In Tel Aviv, Lewis and his staff started to notice that Osiraq had dropped out of the Israeli press—which was either a reason to relax or a reason to worry. Meanwhile, the new administration's Middle East policymakers found themselves preoccupied for months on end with the eruption of a crisis over the introduction of Syrian missiles into Lebanon's Bekaa Valley. Begin warned that Israel might have to take the missiles out, which seemed likely to mean another Syrian-Israeli war. Worse, Begin alleged that Soviet advisers had accompanied Syrian tank crews moving into Lebanon.[16] U.S. special envoy Philip Habib began shuttling around the region, and the larger the Bekaa Valley loomed, the more Osiraq receded. "Honestly, I lost track of the [Osiraq] issue," Lewis noted. "Nobody had asked me to do anything about it. It wasn't in the newspapers. We were really preoccupied with trying to head off war with Syria."[17]

Meanwhile, behind the scenes, Begin concluded that U.S. diplomatic efforts were not going to get rid of the Iraqi reactor. As the Israeli Air Force began honing its plans for the raid, reportedly using "U.S.-supplied satellite photographs, acquired through regular channels," to help with navigation, the U.S. Intelligence Community seems to have been in the dark.[18] "All this was going on without any U.S. intelligence knowledge, as far as I know," Lewis said. "So far as I know, our military attachés didn't know a damn thing." The ambassador would occasionally pause amid the bustle of his daily rounds to think about the Osiraq issue, which he always figured would blow up one day, but in the absence of a sustained push for action from Washington and in the presence of pressing daily demands, the reactor fell largely off the radar of even the most regular interlocutor between the U.S. and Israeli governments.

---

[16] Trudy Rubin, "Habib Pushing Syrian Missile Crisis onto Wider Stage," *Christian Science Monitor*, May 26, 1981.

[17] Lewis, 2012.

[18] Schoenbaum, 1993, p. 276.

In early June, the Reagan administration was trying to shore up the Israel-Egypt Peace Treaty, which was foundering over the mutual suspicion between Begin and Egyptian President Anwar al-Sadat. After 18 months of frosty silence, the two leaders finally broke the ice in a day-long summit meeting in the Sinai on May 28—just days before the Osiraq raid.

Begin not only authorized the attack; he did so with secrecy at its core. Fearing leaks, Begin kept plans for the raid very closely held. Begin quietly summoned his cabinet to his Jerusalem home at 5 p.m. on Sunday, June 7, leaving many of the ministers under the impression they had been invited over for a heart-to-heart talk rather than a cabinet meeting. The cabinet launched into an impromptu session as military updates continued to roll in. Shortly before 7 p.m., the IDF reported that all Israeli planes had returned intact.

The night of the raid, Lewis was at a Tel Aviv hotel, briefing Representative Jack Kemp and a visiting American businessman about the Israeli economy. Begin's office tracked the ambassador down in the businessman's hotel suite, and only then did Israel notify the United States about the Osiraq raid. "Sam, I want you to get ahold of President Reagan immediately," Lewis recalls the prime minister saying over the line. "Our warplanes have destroyed the reactor in Baghdad, and all of our planes have returned safely."[19] "You don't say," Lewis replied, deadpan.[20] The ambassador quickly digested the news and arranged to send an urgent flash cable back to Washington. His own staff was taken entirely unawares. "Everyone here blew up in amazement and shock," Lewis recalled.

When Lewis's bulletin reached the White House, Reagan and National Security Adviser Richard Allen were "thunderstruck."[21] U.S. leaders had not been informed even though U.S. technology had been used. "Reagan never dreamed Begin would do anything of this magni-

[19] Lewis, 2012.

[20] "Attack—and Fallout," 1981, p. 28.

[21] Peter Jessup, interview with Samuel Lewis, Association for Diplomatic Studies and Training, Foreign Affairs Oral History Project, August 9, 1998.

tude without consulting him—not that Reagan would necessarily have been averse to it," Lewis said.[22]

Washington was hardly alone in its ire and surprise. While the world had cheered the July 1976 Israeli raid that freed the hostages being held by terrorists at Entebbe, Uganda, international reaction to another bold Israeli strike five years later was far more mixed. The French foreign minister, embarrassed at having had a French-built facility destroyed, warned that Israel's raid did not serve "the cause of peace in the area." An irate Mitterrand got word to Israel's leadership that they would have been wise to have given him time to reposition France's Iraq policy rather than acting unilaterally. British Prime Minister Margaret Thatcher was far sharper: "Armed attack in such circumstances cannot be justified," she declared. "It represents a grave breach of international law." And the Soviet response was harsher still; the Soviet news agency TASS called the raid an "act of gangsterism" launched with U.S. complicity.[23]

Nor was Reagan the only leader surprised; Sadat had met with Begin just three days before the raid, which earned the Egyptian leader an uncomfortable round of accusations that he had been complicit in the attack.[24] Several conservative Arab leaders were quietly pleased to see Saddam Hussein taken down a peg, but Sadat's inner circle remained convinced that Begin had set them up.[25]

Begin gave no public sign of being concerned. "Israel has nothing to apologize for," he told a Jerusalem press conference on June 9. In less than two weeks, Begin said, Israel would have been unable to "do anything whatsoever in order to prevent the Iraqi tyrant from developing, at least in the near future, between three and five Hiroshima-type nuclear bombs."[26] In an open letter to American Jews and Christians

---

[22] Lewis, 2012.

[23] "Attack—and Fallout," 1981, p. 28.

[24] Bernard Reich, ed., *An Historical Encyclopedia of the Arab-Israeli Conflict*, Westport, Conn.: Greenwood, 1996, p. 390.

[25] Lewis, 2012.

[26] Israel Government Press Office, "Press Conference with Prime Minister Begin, Chief-of-Staff Eitan, Air Force Commander Ivri, and Director of Military Intelligence Saguy," *Press*

on June 12, Begin urged them, "as a free man to free men: do not permit 'punitive' action against Israel because of the deed it was compelled to undertake to save its own life."[27]

The Reagan administration's response intermingled shock, anger, and sympathy—and it was, at least in part, quite distinctly punitive. The administration formally "condemned the attack and then suspended 'for the time being' the delivery of four additional F-16s that were ready to be shipped last week from Fort Worth to Israel."[28] On June 11, Haig reported to the Senate Foreign Relations Committee that the Osiraq raid might represent "a substantial violation of the 1952 agreement" in which Israel assured the United States that it would use U.S. equipment only for self-defense.[29] Reagan's White House not only backed a UN Security Council resolution condemning the raid; it let UN Ambassador Jeane Kirkpatrick consult her Iraqi counterpart while drafting the text.[30] Many in Congress, traditionally the center of gravity for pro-Israel sentiment in the U.S. government, were unamused this time; when Begin later testified before the Senate Foreign Relations Committee to urge it to oppose the proposed sales of Airborne Warning and Control System (AWACS) planes to Saudi Arabia, he was met with a distinctly testy reception.[31]

The Osiraq raid put Reagan in a tricky spot. Reagan had considerable personal sympathy for Israel, sometimes infused with biblical overtones. But he also sometimes bristled at Begin's obduracy and failure to consult—both of which were on display during the Osiraq episode. "Boy, that guy makes it hard for you to be his friend," Reagan

---

Bulletin, June 9, 1981; provided courtesy of the Menachem Begin Heritage Center. I am grateful to Rami Shtivi, the Center's chief archivist, for his kind assistance.

[27] Menachem Begin, open letter to American Jews and Christians, Menachem Begin Heritage Center archives, June 12, 1981.

[28] "Attack—and Fallout," 1981, p. 28.

[29] Alexander Haig, letter to Senate Foreign Relations Committee Chairman Charles Percy, Menachem Begin Heritage Center archives, June 11, 1981.

[30] Sachar, 1987, p. 128; Christa Case Bryant, "Obama-Netanyahu Tensions: Not as Bad as 5 Other US-Israel Low Points," *Christian Science Monitor*, undated.

[31] Sachar, 1987, p. 128.

sometimes grumbled to his aides.[32] Beyond this, Reagan and his team saw considerable strategic benefit in the war that Saddam Hussein had started against Ayatollah Khomeini's Iran, which, with memories of the hostage crisis still smarting and raw, remained America's regional nemesis. As the Israeli historian and diplomat Michael Oren notes, Reagan was "eager to dispel any semblance of collusion in an attack against America's new de facto ally," Iraq.[33]

Reagan's team was split over how harshly to respond. Kirkpatrick was said to be personally sympathetic to the raid. So was Haig, her boss at the State Department, whom associates say had come to conclude that U.S. diplomacy was not going to shut Osiraq down. As such, both Haig and Kirkpatrick wound up agreeing that the United States should take control of any action in the Security Council, rather than letting hotter heads prevail at Turtle Bay and push the United States toward a veto of an overly strong resolution. Not all of their colleagues agreed. As so often happened during this period, the secretaries of state and defense were in opposite corners. "[Secretary of Defense Caspar] Weinberger wanted to come down [on Israel] like a ton of bricks," Lewis recalls. "He was all for real sanctions."[34] (Weinberger does not mention the Osiraq episode in his memoir of his tenure as defense secretary.)[35]

Perhaps in part because of the differing views on his team, and certainly because of the markedly lower strategic stakes for the United States, Reagan was far less angry than Eisenhower and Dulles had been about Israeli action. Unlike Eisenhower and Dulles, who formed a mutually reinforcing circle of outrage at British, French, and Israeli adventurism, Reagan found confirmation among some of his team for some of his own more sanguine inclinations about the raid. Unlike Eisenhower in 1956 and Johnson in 1967, Reagan was confronted with a one-off raid rather than a regional war of massive Cold War conse-

---

[32] Spiegel, 1986, p. 406.

[33] Michael Oren, *Power, Faith, and Fantasy: America in the Middle East, 1776 to the Present*, New York: Norton, 2007, p. 552.

[34] Lewis, 2012.

[35] Caspar Weinberger, *Fighting for Peace: Seven Critical Years in the Pentagon*, New York: Warner, 1991.

quence. Reagan was "not alarmed" by the Osiraq raid, his biographer reports, despite his administration's formal condemnations. At the time, White House officials also reported that "Reagan sympathized" with Begin's view that Iraq's nuclear arms capacity had to be eliminated.[36] (Reagan proved far less sympathetic to Begin's unilateral military moves in Lebanon in 1982, leading to one of the worst confrontations in the history of U.S.-Israel relations.)

Reagan's initial anger that Begin would bomb Osiraq without consulting with Washington also faded fairly quickly. NSC officials relayed to Lewis in Tel Aviv that Reagan was "extremely angry" with Begin for catching him off guard. Lewis responded with a lengthy telegram that recapitulated U.S.-Israeli discussions over Osiraq, and he believes that Reagan's anger cooled fairly shortly after seeing the cable and grasping the background. At a press conference on June 17, Reagan was decidedly even-handed in his remarks. On the one hand, he again "condemned" the raid, said that Israel should have considered other options, and warned that the raid "did appear to be a violation of the law regarding American weapons that were sold for defensive purposes." On the other hand, he added that Israel had ample "reason for concern" about the nuclear ambitions of a hostile neighbor that refused to "even recognize the existence of Israel as a country." Israel, he said, "might have sincerely believed" that the raid "was a defensive move."[37] By the time Begin made his first visit to Washington as prime minister, Reagan was again cordiality itself.

Richard Allen, Reagan's national security adviser, has written that the Reagan administration was genuinely and wholly taken by surprise—and that the president's personal feelings about the raid were not as censorious as the official reaction he authorized. Allen recalls sitting on his back porch in Arlington, Virginia, churning through paperwork, when the White House Situation Room abruptly called him. From a secure phone line in his basement, Allen was told by a

[36] Lou Cannon, *President Reagan: The Role of a Lifetime*, New York: PublicAffairs, 2000, p. 341.

[37] Ronald Reagan, statement at press conference, Menachem Begin Heritage Center archives, June 17, 1981.

duty officer that Israeli F-16s and F-15s were streaking away from a raid on Iraq's reactor. As U.S. officials scrambled to figure out what had happened (including the mystery of how the F-16s had managed to fly so far beyond their usual range), Allen asked the White House switchboard to put him through to President Reagan—who was in the process of boarding his helicopter. A somewhat annoyed Reagan was brought to the phone, and with chopper blades roaring in the background, listened to Allen's update. "Why do you suppose they did that?" Reagan asked. Allen said he assumed the Israelis had not wanted to risk waiting for Osiraq to come online. After a pause, Reagan reportedly said, "Boys will be boys!"[38]

Other national security principals were not as amused. Haig was at home when he was told of the raid, and he was as surprised as anyone else—and worried about the awkwardness of Israel's use of U.S.-made aircraft and perhaps munitions in an offensive raid. Haig began scribbling on a yellow legal pad a draft statement condemning the raid. "Israel's action had been shocking, and there would be consequences," Haig later wrote. "Not only could the United States not condone the raid, it would have to take some action against Israel."[39] But even the irked secretary of state admitted to mixed emotions and to wondering whether history might look rather kindly on Begin's decision to kick the bomb away from the grasping hands of Saddam Hussein.

In a noisy Oval Office meeting directly after the raid, Vice President George H. W. Bush, White House Chief of Staff James Baker, and longtime Reagan aide Michael Deaver "argued strongly for punitive actions against Israel, including taking back aircraft and delaying or canceling scheduled deliveries." Allen recalls Defense Secretary Weinberger as "angry but measured" and Director of Central Intelligence William Casey as studiously mute. Secretary of State Haig told Allen that he sympathized with Israel but was under pressure from his department and U.S. partners to condemn the attack, which ultimately

[38] Richard V. Allen, "Reagan's Secure Line," *New York Times*, June 7, 2010, p. A23.

[39] Alexander Haig, *Caveat: Realism, Reagan, and Foreign Policy*, New York: Macmillan, 1984, pp. 182–183.

led him to clear on U.S. criticism of the attack. According to Allen, "The president himself said little, listening patiently."[40]

Reagan's own diaries paint a somewhat different portrait of the episode. His first reaction was downright biblical: "Got word [of] Israeli bombing of Iraq - nuclear reactor. I swear Armageddon is near." After returning to the White House, Reagan noted, "P.M. Begin informed us after the fact." In a conversation with the president later that day, Begin argued that waiting to mount the raid until after further French uranium shipments could have meant unleashing radiation over Baghdad. Reagan recalled being dunned by five Arab ambassadors to the UN for being too soft on Israel and then having the Israeli ambassador protest "the harsh action we'd taken." Reagan replied that his administration would "keep on trying to bring peace to Lebanon & then Between Arabs & Israelis." In June, Reagan complained to his diaries that the outgoing Carter administration had failed to warn his team during the transition of the looming threat posed by Osiraq. "Amb. Lewis cabled word to us after the Israeli attack on Iraq," an exasperated Reagan wrote, and "now we find there was a stack of cables & memos tucked away in St[ate] Dept. files."[41] This tracks rather well with Lewis's recollection that the transition material about Osiraq got lost in the shuffle until the system burped it back out again in the wake of the Israeli raid.

As Haig recalled, reaction inside the administration "combined astonishment with exasperation," with some aides urging punitive sanctions against Israel while Haig and others argued that a show of U.S. disapproval would be wiser than "policies that humiliated and weakened Israel."[42] That balancing act led to the decision to delay the shipment of four more F-16s to Israel to make U.S. disapproval clear and public and to underscore American displeasure over the use to which a previous batch of F-16s had been put. The raid also significantly complicated the administration's efforts to hammer home its controversial

[40] Allen, 2010.

[41] "Reagan Diaries: Carter Failed to Tell Reagan About Osirak," History News Network, May 28, 2007.

[42] Haig, 1984, p. 184.

proposed sale of AWACS planes to Saudi Arabia—the largest single U.S. arms sale to date to an Arab state. By overflying Saudi territory, Israel gave Saudi officials the chance to argue publicly that they clearly needed AWACS technology to protect themselves from Israeli adventurism. (However, as David Schoenbaum has dryly noted, the "U.S. AWACS already in Saudi Arabia failed to detect the Israeli planes.")[43]

Reagan also gave Kirkpatrick license to shape and pass a Security Council resolution deploring the Israeli raid—and even to consult with Saddam's UN ambassador while doing so. Radical Arab states were pushing for a conspicuously and provocatively tough Security Council resolution "that would boycott or expel Israel from the UN," which its proponents presumed would be met with a veto from Washington, further exacerbating tensions between the United States and the Arab world.[44] The Reagan team decided that it would be a mistake to sit back and let hotheads set the tone. In New York, Kirkpatrick worked to craft a resolution that condemned Israel for the raid but stopped short of calling for sanctions. The administration calculated that leading the diplomacy on the resolution gave them more control, whereas standing back could have opened up a diplomatic vacuum that would have been filled with strident calls for punitive action.

The compromise resolution passed the Security Council unanimously on June 19. It avoided sanctions or talk of expulsion, but otherwise, Resolution 487 contained little succor for Israel: the Security Council "strongly condemns" Israel's raid, which it finds "in clear violation of the Charter of the United Nations and the norms of international conduct," "calls upon Israel to refrain in the future from any such acts or threats thereof," calls the raid "a serious threat to the entire IAEA [International Atomic Energy Agency] safeguards regime" that undergirds the Nuclear Nonproliferation Treaty, "fully recognizes" Iraq's "inalienable sovereign right" to peaceful nuclear development, calls on Israel "urgently to place its nuclear facilities under IAEA safeguards," and considers Iraq "entitled to appropriate redress for the

[43] Schoenbaum, 1993, p. 277.

[44] Herzog and Gazit, 2005, p. 345.

destruction it has suffered" at Israel's hands.[45] This was a jagged pill for Begin to swallow. But Reagan was pleased enough with Kirkpatrick's work in New York to take time to call her with his personal congratulations.[46]

If the Reagan administration thought its response supple enough to send a series of careful signals, the Begin government seems not to have been overly appreciative. A fuming Begin later told a Knesset committee that the United States "had given him a document that supported his suspicions that Iraq was indeed planning to build a bomb." Israeli officials later conceded that Begin had overstated the case; as one senior Israeli official put it, "The aim of the [U.S.] paper was to play down the possible danger of the reactor."[47] Begin may have been annoyed by the ongoing divergence in U.S. and Israeli intelligence assessments of Osiraq. In any event, he was hardly solicitous of U.S. feelings after the raid. When he heard of a (false) news story reporting that Weinberger had proposed slashing U.S. military aid to Israel after the raid, he called it "chutzpah" in private—and was far harsher in public. "By what morality were you acting, Mr. American Secretary of Defense?" he asked on the stump. "Haven't you heard of 1.5 million little Jewish children who were thrown into the gas chambers?"[48] The Pentagon gently made clear that Begin had been misinformed. But it was clear that the prime minister's hackles had been raised by the administration response—and that he may have also thought that teeing off on Weinberger and the Reagan administration made for a winning electoral-season ploy.

Years later, the Reagan administration's response to being surprised in June 1981 clearly emerges as far less angry than its Republican predecessors' in 1956. Reagan himself fairly quickly got over his initial shock that Begin would take so major a step without consulting him. But on balance, the Reagan administration's response had a distinctly

---

45 United Nations Security Council, Resolution 487, June 19, 1981.

46 Cannon, 2000, p. 164.

47 "Long Shadow of the Reactor," *TIME*, Vol. 118, No. 1, July 6, 1981, p. 35.

48 "Attack—and Fallout," 1981, p. 28.

sharp tone. As Steven Spiegel wrote of the F-16 delay, it was "unprecedented" to suspend delivery of weapons already sold to Israel. "The last time any such action had been taken by an administration was Eisenhower's temporary suspension of economic aid in October 1953," Spiegel notes. "Even after the shipments were resumed, the president's attitude was increasingly negative, and tensions between Washington and Jerusalem continued."[49] These tensions worsened dramatically during the U.S.-Israeli confrontation over Begin and Ariel Sharon's 1982 invasion of Lebanon, which led to some of the frostiest moments in the history of the U.S.-Israel special relationship.

Part of the reason for the American surprise was simply Begin's own understanding of the stakes. When the prime minister explained the raid to the Israeli public, he used the starkest of terms: "We chose this moment: now, not later, because later may be too late, perhaps forever. And if we stood by idly, two, three years, at the most four years, and Saddam Hussein would have produced his three, four, five bombs... another Holocaust would have happened in the history of the Jewish people."[50] Framed in those terms, the intermittent and relatively low-priority U.S. diplomacy with France and other Iraqi nuclear suppliers never had much chance of success unless infused with great urgency and high-level effort.

Israeli leaders considered the attack on Osiraq not so much 1967-style preemption as "anticipatory self-defense."[51] While the reactor was poised to go online, some U.S. officials concluded that Iraq was years away from producing nuclear weapons. Some subsequent studies have tended to back up the more sanguine U.S. analysis rather than the more worried Israeli one. The former Pentagon official Colin Kahl has argued that even if Saddam chose to use Osiraq to produce atomic bombs, the reactor was inefficient enough that "it would still have taken several years—perhaps well into the 1990s—to produce enough plutonium for a single bomb." Kahl adds that the chances of a swift

---

[49] Spiegel, 1986, p. 409.

[50] Colin Kahl, "Before Attacking Iran, Israel Should Learn from Its 1981 Strike on Iraq," *The Washington Post*, March 2, 2012.

[51] Pollack, 2002, p. 369.

Iraqi breakout "were further undercut by the presence of French technicians at Osirak, as well as regular inspections by the International Atomic Energy Agency. As a result, any significant diversion of highly enriched uranium fuel or attempts to produce fissionable plutonium would probably have been detected."[52]

Perhaps because it failed to fully assimilate the depth of Begin's foreboding and to fully absorb the gist of Israel's intelligence analysis, the Reagan administration wound up being startled by the Osiraq raid. Still, as David Schoenbaum points out, "the surprise itself was surprising."[53] If Israel had been carefully opaque about the imminence of military action, it had been publicly blunt about the extent of its worry. Then again, Ben-Gurion had been quite clear about his anxieties about Nasser in 1956. Part of the Reagan administration's ire in 1981 stemmed precisely from the effort and investment that successive U.S. governments had put into a closer, more consultative relationship with Israel—and from the irritation that senior administration officials, including President Reagan himself, felt at Begin's insistence on often steering his own course without paying overmuch heed to U.S. equities or cautions.

Lewis suspects that if Begin had asked Reagan for permission to attack Osiraq, the Israeli leader would have been told, in effect: "We don't want you to do it, but you have to make your own decisions"—a message, Lewis notes, that is not entirely unlike some recent exchanges over Iran between President Barack Obama and Prime Minister Binyamin Netanyahu. But Lewis's boss, Secretary of State Haig, had a harsher view: He remembers Israeli officials telling him shortly after the raid that they had "agonized" about whether to give Washington advance notice but ultimately decided against telling the United States because it would have told Israel not to bomb. "In this judgment," Haig wrote flatly, "they were correct."[54]

---

[52] Kahl, 2012.

[53] Schoenbaum, 1993, p. 277.

[54] Haig, 1984, p. 182.

Case IV: The Raid on al-Kibar, 2007

On September 6, 2007, Israeli F-15s again raced into the air space of a radical Arab neighbor to take out a nuclear reactor. This time, the state was Syria, and the reactor was housed not in the outskirts of its capital but in the small village of al-Kibar, nestled along the Euphrates. In 1981, the Israeli code name had been Operation Opera; in 2007, according to press accounts, it was Operation Orchard.

There are considerable similarities between the 1981 and 2007 cases. But there is at least one major difference between Osiraq and Kibar: While Menachem Begin had taken the Reagan administration by surprise in 1981, Ehud Olmert made a point in 2007 of not only consulting with the United States but of pushing George W. Bush's administration to bomb the reactor itself—as it were, something close to a nonproliferation first right of refusal. Olmert had resolved that if the United States would not destroy the facility, Israel would take care of the problem itself.[1] Both Begin and Olmert shared Israel's mounting alarm about the nuclear facilities in their gunsights, and both prime ministers did not (at least as far as we know) directly warn the United States when Israel's strike was imminent. But Olmert took pains to telegraph Israeli action in a way that Begin never came close to doing—making it clear to the Bush administration that Israel was highly likely to resort to bombing if an already overcommitted United States would

[1] Dick Cheney recounts Olmert's warning in his memoir, and it strains credulity to think that Olmert would make so direct and important a statement to the vice president without also sharing something similar with the president. Still, Bush's memoir is more circumspect about any direct warning from Olmert of the use of Israeli force.

not. As one senior U.S. official put it, "We had post-Iraq syndrome, and the Israelis had preemption syndrome."[2]

Washington had been concerned for some time about Syria's potential nuclear ambitions; in the 1990s, U.S. pressure scotched attempts by Syria's long-ruling Baathist dictator, Hafiz al-Asad, to buy nuclear reactors from Russia and Argentina, supposedly for non-military research. Starting around 2001, U.S. satellites looked on as Syrian engineers began to build an odd facility in a remote Euphrates Valley town. In March 2007, operatives from Israel's Mossad intelligence service raided the Vienna home of the head of Syria's Atomic Energy Commission. They managed to recover from his computer "roughly three dozen color photographs taken from inside the Syrian building, indicating that it was a top-secret plutonium nuclear reactor." On March 8, Mossad chief Meir Dagan briefed Olmert on the raid. "I knew from that moment, nothing would be the same again," Olmert later said. "The weight of this thing, at the existential level, was of an unprecedented scale."[3] Like Begin, Olmert concluded that Israel could not wait for the reactor to go hot before destroying it, lest an Israeli strike spread deadly radiation and invite international calumny. Olmert swiftly concluded that the reactor would have to be destroyed.

From the outset, Olmert was particularly concerned about operational security. He convened a small planning group, including Defense Minister Amir Peretz (who would be replaced by Ehud Barak by the time of the Kibar raid), IDF Chief of Staff Gabi Ashkenazi, Dagan, and other Israeli intelligence leaders. They often met at Olmert's official residence to avoid attracting attention from curious onlookers watching the traffic into and out of the prime minister's office. Olmert also quietly briefed former Israeli prime ministers Shimon Peres, Ehud Barak, and Binyamin Netanyahu.[4]

As David Sanger has reported, the Kibar project lacked army guards or suspicious barbed-wire enclosures—nothing, in other words,

---

[2] David Sanger, *The Inheritance: The World Obama Confronts and the Challenges to American Power*, New York: Three Rivers, 2009, p. 278.

[3] David Makovsky, "The Silent Strike," *The New Yorker*, September 17, 2012, p. 34.

[4] Makovsky, 2012, p. 35.

obvious enough to catch the eye of an overhead onlooker. But the watchful Americans continued to harbor their suspicions. Israel seems to have first tipped off the United States on April 18, 2007, when Defense Secretary Robert Gates met with his Israeli counterpart, Peretz, during a routine trip to Israel. Peretz, who is not a fluent English speaker, had a special "index card prepared for him, which he could refer to as he divulged the news about the reactor."[5] Dagan was also in Washington for an urgent meeting.[6] Elliott Abrams, then a deputy national security adviser handling the Middle East, recalls that Olmert had asked to have Dagan brief President Bush personally, but the administration fobbed him off with a meeting with National Security Adviser Stephen Hadley and Abrams. Abrams adds that Vice President Cheney sat in on the briefing in Hadley's office.[7]

Dagan came armed with considerably stronger intelligence than the United States had been able to amass from orbit. Rather than satellite imagery, he carried the Mossad's photographs from inside the Kibar reactor. Some of the photos had been taken in 2003 and 2004, "apparently by a Syrian who had been 'turned,' or paid handsomely for his snapshot collection." Hadley reportedly noted that the Kibar facility lacked the power lines one would expect to see sprouting forth from a civilian nuclear reactor producing electricity. Dagan also showed the Americans one particularly prized photo showing the head of the Syrian Atomic Energy Commission standing close to the facility—along with a North Korean nuclear official who worked at North Korea's Yongbyon nuclear site. One U.S. official who looked at the Kibar images declared the Syrian facility a "carbon copy" of Yongbyon.[8]

Hadley later told reporters that he immediately expected that Olmert "would demand that the United States destroy this reactor in the desert, or stand back while Israel took care of the problem itself." Dagan briefed CIA Director Michael Hayden on Israel's photographic

---

5 Makovsky, 2012, p. 35.

6 Sanger, 2009, p. 271.

7 Elliott Abrams, "Bombing the Syrian Reactor: The Untold Story," *Commentary*, February 2013.

8 Sanger, 2009, pp. 271–272.

evidence, and Hayden in turn briefed President Bush. Bush was shown the Israeli photos, and he later noted that his team strongly suspected that "we had just caught Syria red-handed trying to develop a nuclear-weapons capability with North Korea."[9] The president is said to have swiftly asked Hadley and Secretary of State Condoleezza Rice to urge the Israelis not to attack Kibar.[10] "The disastrous intelligence failure on weapons of mass destruction in Iraq was fresh in everyone's mind," and Bush ordered Hayden and the Intelligence Community to confirm that Israel had the intelligence right this time. "Gotta be secret," the president reportedly said, "and gotta be sure."[11] Shortly after being shown the Israeli photos, Bush and Olmert spoke by phone. "George, I'm asking you to bomb the compound," Olmert said. "Thank you for raising this matter," Bush replied. "Give me some time to look at the intelligence, and I'll give you an answer."[12]

Hayden tasked a group of CIA officials to assess the quality of the intelligence about the Kibar reactor, including the Syria-North Korea connection. They spent months poring over the data.[13] The Israeli snapshots from Kibar were run past the National Geospatial-Intelligence Agency (NGA) and a small group of analysts at the Department of Energy.[14] But in the main, the CIA-led effort was closely compartmentalized, with congressional notifications kept to a minimum and knowledge of the reactor limited even within the Intelligence Community. Hayden worried that a leak might put Asad "in a position where he felt publicly humiliated and thought he had to respond if the facility were attacked" by the United States or Israel.[15]

---

[9] George W. Bush, *Decision Points*, New York: Crown, 2010, p. 421.

[10] Sanger, 2009, p. 275.

[11] Makovsky, 2012, p. 35.

[12] Bush, 2010, p. 421.

[13] Bob Woodward, "In Cheney's Memoir, It's Clear Iraq's Lessons Didn't Sink In," *Washington Post*, September 11, 2011.

[14] Makovsky, 2012, p. 36.

[15] Michael Hayden, "Correcting the Record About that Syrian Nuclear Reactor," *Washington Post*, September 22, 2011.

Meanwhile, Hadley directed a subset of the Deputies Committee—known as the Drafting Committee—to work through possible U.S. policy options. The committee had to leave its documents in the Situation Room for greater security. Its members reportedly included Elliott Abrams of the NSC staff, Eric Edelman from the Defense Department, and Eliot Cohen and James Jeffrey from the State Department.[16] Abrams, who considers the process an exemplar of policymaking rigor, recalls an extraordinarily closely held process, with discussion papers "kept under lock and key" and executive assistants "kept out of the loop."[17]

Israel's request that the United States destroy the Syrian facility triggered a rolling policy debate within the administration. The discussions quickly came to center not in the subcabinet-level Drafting Committee but in weekly Small Group meetings chaired by Hadley, with participants including Cheney, Rice, Gates, Hayden, Joint Chiefs of Staff Chairman Peter Pace, and Director of National Intelligence (DNI) Mike McConnell. Abrams writes that Small Group principals several times "trooped over to the president's living room in the residence section of the White House to have it out before him, answer his questions, and see what additional information he sought."[18] Rice recalls principal-level discussion of the Israeli request to take out the reactor taking up "the better part of two months."[19]

Abrams, who was the notetaker for these sessions, remembers a clear array of options: "overt or covert, Israel or United States, military or diplomatic."[20] Bush recalls his team considering three key options: a bombing raid, which would be simple militarily but "would create severe blowback"; a covert raid, which both the CIA and the Defense Department ultimately wrote off as "too risky"; or an international push to expose the Syrian project and have it dismantled under the

---

16 Makovsky, 2012, p. 36.

17 Abrams, 2013.

18 Abrams, 2013.

19 Condoleezza Rice, *No Higher Honor: A Memoir of My Years in Washington*, New York: Crown, 2011, p. 708.

20 Abrams, 2013.

watchful eyes of the IAEA.[21] Bush's memoir does not mention an obvious fourth option that might have been offered for principal-level consideration: unleashing Israel to take care of the problem itself.

Bush is said to have "seriously considered ordering an American military strike on the reactor," which some senior U.S. officials thought would be less likely to set off a regional conflagration than an Israeli airstrike.[22] General Pace told the president that bombing Kibar would not pose "much of a military challenge," according to Abrams, and the interagency "developed elaborate scenarios for U.S. and Israeli military action."[23]

Cheney, by all accounts, was the most vociferous proponent of a U.S. strike to take out Kibar. He argued that it would cripple Syria's incipient nuclear ambitions, warn North Korea off of further proliferation, and deter Iran from moving ahead with its own nuclear program. The vice president, who admired Israel's 1981 raid on Osiraq, saw Kibar as an important test of U.S. resolve and seriousness, and he urged his fellow Small Group members to use a strike to send a clear signal that Washington would not tolerate "the proliferation of nuclear technology to terrorist states."[24] As a bonus, he argued, such a strike would increase the chances of finding a diplomatic outcome to the Iranian and North Korean nuclear standoffs simply by reminding them that U.S. force in the service of nonproliferation remained very much on the table. Cheney added a layer of tactical argument too, noting that the Kibar reactor was an obvious target for American airpower: remote, far removed from Syrian cities, and isolated amid miles of desert.

Others in the Small Group disagreed sharply with Cheney, including Gates and Rice.[25] With the counterinsurgency "surge" under way in Iraq and combat continuing in Afghanistan, few were eager to

---

[21] Bush, 2010, p. 421.

[22] Sanger, 2009, p. 276.

[23] Abrams, 2013.

[24] Dick Cheney with Liz Cheney, *In My Time: A Personal and Political Memoir*, New York: Threshold, 2011, p. 468.

[25] Rice, 2011, p. 708.

unleash a chain of events that could lead to another war in the Middle East. Beyond that, Hadley told a *New York Times* reporter that he did not think the Kibar facility met "the standards of the 'Bush Doctrine' for a preemptive strike," noting that CIA analysts did not think the United States could prove that Syria planned to acquire nuclear weapons at Kibar.[26] The disastrous U.S. intelligence failure on Iraq's supposed nuclear arms program clearly hung heavy in the air. Gates dryly told an aide, "Every administration gets one preemptive war against a Muslim country, and this administration has already done one."[27] With memories of the ill-fated 2006 Israeli war against Hizballah still fresh in Rice's mind, the secretary of state worried that the Israeli military might not be up to the task and that "a strike would lead to a wider conflict, including war with both Syria and Hezbollah."[28] Bush too had "lost ... confidence" in Israel during the recent Lebanon war.[29]

Cheney recalls his opponents warning that a strike on Kibar—whether launched by Israel or the United States—could spark a wider Middle East war, presumably drawing in not just Israel and Syria but also Hizballah, Hamas, and perhaps even Iran. Cheney adds that some Small Group members fretted that Syria could retaliate against U.S. troops in Iraq. The vice president writes that he brushed these objections aside, downplaying the likelihood of a regional conflict while arguing that the riskiest course of action would be letting the Pyongyang-Damascus nuclear collaboration remain intact. Cheney laments that the disastrous intelligence failure that assessed that Saddam Hussein had an active nuclear program in the run-up to the 2003 U.S. invasion had "made some key policymakers very reluctant to consider robust options for dealing with the Syrian plant."[30] For his part, the hawkish Abrams disliked the option of hauling the Syrians in front of the IAEA and the Security Council, arguing that Israel would never

[26] Sanger, 2009, p. 276.

[27] Makovsky, 2012.

[28] Makovsky, 2012, p. 36.

[29] Bush, 2010, p. 422.

[30] Cheney, 2011, p. 469.

trust the UN, that publicly airing the issue could foreclose Israeli military action, that IAEA Director General Muhammad al-Baradei would cave in to Syrian pressure, and that "an issue of this importance should be handled in the White House" and not handed over to the State Department.[31]

Among those reluctant to act was President Bush himself. In his memoirs, Bush makes clear his apprehensions about a regional backlash from a U.S. bombing campaign.[32] Cheney recalls raising the Kibar issue with the president at one of their private weekly lunches, on June 14, and arguing that the administration should take an "aggressive strategy" toward Syria and Iran's attempts to develop nuclear-arms capabilities. Bush was not persuaded to launch a unilateral U.S. strike on the Syrian reactor but did agree to convene his national security principals in the Yellow Oval Room of the White House on Sunday evening, June 17.[33]

Cheney recalls DNI McConnell telling the meeting that the Intelligence Community had "high confidence" that the Kibar facility was a reactor. "It's about as good as it gets," McConnell reportedly added. Cheney argued in favor of a U.S. airstrike to destroy the reactor. Other principals argued instead for going public with U.S. assessments of Kibar, taking the matter to the UN Security Council and the IAEA, and seeking a diplomatic route to pressure the Syrians to abandon their nuclear program.[34]

The very highest levels of the U.S. government heard directly from Olmert days later. The Israeli prime minister joined Bush and Cheney for a lunch meeting on June 19. That evening, Cheney recalls, Olmert bluntly urged the administration over dinner at Blair House to "take military action to destroy the facility" and made it clear that "Israel would act if we did not."[35] Olmert is said to have argued that a U.S.

---

[31] Abrams, 2013.

[32] Bush, 2010, p. 421.

[33] Cheney, 2011, pp. 469–470.

[34] Cheney, 2011, pp. 470–471.

[35] Cheney, 2011, p. 471.

strike on Kibar would "kill two birds with one stone"—both setting back the nuclear program of a notoriously abusive regime and deterring Iran. And he was blunt about the consequences of U.S. inaction: if the United States was not willing to take out the reactor, Israel was.[36]

According to Bob Woodward of *The Washington Post*, Hayden later told colleagues that he had deliberately tried to underscore the weakness of the intelligence case to reduce the chances of a U.S. attack on a facility that the CIA was far from convinced was military in nature. (Hayden has disputed this, writing that "the debate in the U.S. government over [the reactor's] fate was informed by hard facts" without prejudging the decisions later made on that empirical basis.)[37] CIA analysts cautioned that they had found "no evidence of plutonium reprocessing capability at the site" or its surrounding environs and had not been able to identify any way "to manufacture uranium fuel."[38]

Cheney recounts that he added a new argument to his ongoing case for U.S. strikes: they would "enhance our credibility" in the Middle East, "taking us back to where we were in 2003, after we had taken down the Taliban, taken down Saddam's regime, and gotten [Libyan dictator Muammar] Qaddafi to turn over his nuclear program." He found no takers. "Does anyone here agree with the vice president?" Bush asked after Cheney had finished speaking. The vice president remembers looking around the table and seeing no supporters. (Abrams says he did not raise his hand because he wanted the Israelis to take out the reactor themselves and restore their military deterrent, which had eroded after the 2006 war with Hizballah.)[39] Gates and Rice both preferred the diplomatic route, and Abrams later wrote that the defense secretary "argued for preventing Israel from bombing the reactor and urged putting the whole relationship between the United States and Israel on the line."[40] (None of the currently avail-

---

[36] Makovsky, 2012, p. 36.

[37] Michael Hayden, 2011.

[38] Woodward, 2011.

[39] Abrams, 2013.

[40] Abrams, 2013.

able accounts by the principals includes this.) Cheney concluded that the president too had chosen diplomacy rather than force. According to Woodward, after Cheney made his case, "Bush rolled his eyes."[41] The president finally "decided that he could not order another military strike on a state he accused of possessing a program to build weapons of mass destruction."[42]

The full NSC met again on July 12. This time, Bush "declared that he would send an envoy to Syria with an ultimatum for Assad to begin dismantling the reactor," overseen by the five permanent members of the UN Security Council.[43] That left the United States committed to a diplomatic path—and the prospect of an Israeli strike hanging over the principals' heads. During this period, Bush reportedly asked Rice how Israel was likely to react to hearing the U.S. decision. The secretary of state is said to have replied that Olmert's government "would go along with the idea of taking the information to the United Nations [Security Council] and working for multilateral action to shut down the facility." An unhappy Cheney says he flatly disagreed, invoking the memory of Osiraq and warning to expect Israel to take matters into its own hands.[44] Rice says she worked to come up with "an alternative plan that involved going rapidly to the United Nations, exposing the program, and demanding that it be immediately dismantled."[45]

Bush broke the news to Olmert in a secure phone call on the morning of July 13. "I cannot justify an attack on a sovereign nation unless my intelligence agencies stand up and say it's a weapons program," Bush recalls saying on the call.[46] He added that he had settled instead on a diplomatic option through the IAEA and the UN Security Council, "backed up by the threat of force." (Abrams, who was in the Oval Office for the call, remembers the conversation differently from

---

[41] Woodward, 2011.

[42] Sanger, 2009, p. 276.

[43] Makovsky, 2012, p. 37.

[44] Cheney, 2011, p. 472.

[45] Sanger, 2009, pp. 276–277.

[46] Bush, 2010, p. 421.

Bush; he has written that a forceful Olmert said he "cannot accept" Bush's decision and warned that if the United States would not act, Israel would.)[47] The president pushed the Israeli leader about whether his country really wanted to be fingered as the source of the intelligence that spurred another U.S. military strike in the Middle East. Instead, Bush suggested sending Rice to the region. Olmert demurred, arguing that diplomacy was a dead end that would simply let Syria play for time while Israel lost its military option.[48] As Cheney recounts, the disappointed prime minister said that the diplomatic route "wouldn't work for Israel." Olmert reportedly warned that Israel did not have any time to lose and would be forced to deal with Kibar before it "went hot" for fear of unleashing horrifying nuclear fallout in a strike after the reactor was loaded with fuel—precisely the same argument that Begin made about Osiraq. Israel, Olmert is said to have added, could not place its fate "in the hands of the UN or the IAEA."[49] Olmert told Bush that Kibar was an "existential" threat to Israel, calling it "something that hits at the very serious nerves of this country."[50] Olmert worried that U.S. officials who opposed an Israeli strike might take to the press to scuttle it, and Bush assured Olmert that his team would stay "buttoned up." As Makovsky notes, "at no point did Bush suggest that the U.S. would block Israeli action."[51] Abrams recalls Bush listening calmly, hanging up, and admiringly saying, "That guy has guts" (or, reportedly, another anatomical variant of the same thought).[52]

Olmert's inner circle was unhappy about the U.S. decision, which left the Kibar problem back in Israel's lap. Israeli officials complained that the United States was ignoring a nuclear arms program far more advanced than Saddam Hussein's. "It was laughable logic," one senior Israeli official told Sanger. "Whatever happened to the George Bush

[47] Abrams, 2013.

[48] Makovsky, 2012, p. 37.

[49] Cheney, 2011, p. 472.

[50] Bush, 2010, p. 421.

[51] Makovsky, 2012, p. 37.

[52] Abrams, 2013; author interview with expert involved in Kibar policy.

who said that after 9/11, we could not let threats fester?"[53] Bush himself remembers Olmert ending their call by saying, "I must be honest and sincere with you. Your strategy is very disturbing to me."[54]

Behind the scenes at Langley, some of the CIA experts who had worked for months on the Kibar file breathed a sigh of relief. They were pleased enough with the outcome to take the trouble to quietly mint a small number of memorial coins, featuring "a map of Syria with a star" at Kibar on one side—and when one flipped the coin over, the phrase "No core/No war."[55]

Even as Olmert warned that the UN route was a dead end, at least some U.S. officials took a somewhat jaundiced view of his motivations. Olmert's premiership had almost been stillborn when he quickly launched an ambitious 2006 war to root out Hizballah from southern Lebanon. But Olmert's war did more to dent Israel's deterrent than to multiply it; the Shia militia took Israel's worst punch and expanded its influence over Lebanese political life. Meanwhile, Olmert found himself under investigation for alleged campaign finance abuses, and some American officials told reporters that he might have been relieved to have an opportunity to rally the Israeli public around a decisive, morally unambiguous strike on a despised foe seeking doomsday weapons—a sort of Israeli "Wag the Dog" allegation. "They thought this was Osiraq all over again," a senior U.S. official told Sanger.[56] The arguments between Washington and Jerusalem grumbled on over the summer of 2007; it is perhaps some indicator of hard feelings that American officials subsequently ascribed fairly crass political motives to Israel's leader.

On September 1, a top Olmert aide "told the White House that preparations were almost complete" for a raid on Kibar. The Israeli official did not discuss the potential timing of a strike. Israel also tipped off Britain's MI6—and again stayed mum about its timing. Olmert's team, Sanger writes, was "careful not to inform Washington of the pre-

---

[53] Sanger, 2009, p. 277.

[54] Bush, 2010, p. 422.

[55] Woodward, 2011.

[56] Sanger, 2009, p. 277.

cise timing so that both nations could claim, with technical accuracy, that the Americans had not known about the attack in advance."[57] That phrasing does suggest that the Bush administration had been given some form of notice by Israel before the Kibar raid. It certainly suggests that Bush was vastly less surprised in 2007 than Eisenhower in 1956 or Reagan in 1981—far more like Johnson in 1967, who clearly understood that the Israelis were straining at the leash and ruefully expected them to bolt.

Four days later, Olmert convened his security cabinet, which blessed the operation and left its precise timing in the hands of Olmert, Defense Minister Barak, and Foreign Minister Tzipi Livni. "The Israelis did not seek, nor did they get, a green or red light from us," Abrams writes.[58] Just before midnight on September 5, Israel scrambled its F-15s and F-16s, and in the dead of night, they destroyed the reactor.

Olmert took pains to break the news to Bush personally. On September 6, before the U.S.-made planes used in the raid had even landed back in Israel, Olmert called the president from the Israeli defense ministry. "I just want to report to you that something that existed doesn't exist anymore," Olmert said, seemingly mindful that others might be listening in on an insecure line. "It was done with complete success."[59] Bush reportedly offered little by way of reaction, perhaps because of the open international phone line and perhaps because of his own calculations.[60]

As questions mounted after the raid, Israel stayed atypically circumspect. Many Syrians refused to believe that Israel had taken out a Syrian reactor, and Israel quietly briefed Egyptian and Jordanian leaders to tamp down public speculation about the bombing. Olmert personally briefed Russian and Turkish leaders.[61] Cheney notes that Israeli officials asked after the Kibar raid that Washington stay mum about

---

[57] Sanger, 2009, p. 278.

[58] Abrams, 2013.

[59] Makovsky, 2012, pp. 37–38.

[60] David Makovsky, discussion with author, January 16, 2013.

[61] Makovsky, 2012, p. 39.

its intelligence on the facility, fearing that publicity about the Syrian program might further embarrass Syrian President Bashar al-Asad and increase the chances that he would lash back and escalate. The administration agreed.[62] "Neither the Syrian, U.S. nor Israeli governments said much about" the raid, Hayden later wrote, and Asad simply "let the facility's destruction pass."[63] The "central worry," Gates recalled, was that "if you play this wrong there could be a war between Israel and Syria."[64]

That caution went all the way to the top. At a White House press conference on September 20, a reporter asked Bush "whether you supported this bombing raid, and what do you think it does to change the dynamic in an already hot region in terms of Syria and Iran and the dispute with Israel and whether the U.S. could be drawn into any of this?" He refused to comment. When the reporter persisted, Bush shot back, "Saying I'm not going to comment on the matter means I'm not going to comment on the matter."[65] Behind the scenes, Cheney had argued that exposing the Israeli raid would also expose North Korean involvement in Syria's nuclear program. Bush had quietly suggested to Olmert "that we let some time go by and then reveal the operation as a way to isolate the Syrian regime." But Olmert demurred, arguing that anything less than total secrecy "might back Syria into a corner and force Assad to retaliate." Bush decided that he had little choice but to bow to Olmert's wishes. "I kept quiet," he recalls, "even though I thought we were missing an opportunity."[66]

According to Makovsky, an Israeli general later noted, "Olmert said he did not ask Bush for a green light, but Bush did not give Olmert a *red* light. Olmert saw it as green."[67] Bush himself disagreed—up to a point. "Prime Minister Olmert hadn't asked for a green light, and I

---

[62] Cheney, 2011, p. 472.

[63] Hayden, 2011.

[64] Sanger, 2009, p. 270.

[65] George W. Bush, statement at press conference, September 20, 2007.

[66] Bush, 2010, p. 422.

[67] Makovsky, 2012, p. 37; my italics.

hadn't given one," he writes. "He had done what he believed was necessary to protect Israel."[68] And even if the Bush administration had not seen those actions as entirely necessary or requiring a formal blessing, it did not choose to second-guess Israel or offer anything but tactical cooperation after the fact. Nor did the administration's feathers seem particularly ruffled. As Rice herself coolly put it, "Ehud Olmert thanked us for our input but rejected our advice, and the Israelis then expertly did the job themselves."[69]

Rice's sangfroid, even in hindsight, may well be telling. The Bush administration was not willing to flash Israel a green light to strike Kibar, but it did not exactly flash a red one either. If anything, the factor that Rice, Hadley, Gates, and ultimately Bush found most unsettling was the prospect that the United States, already bogged down in costly and bloody wars in Iraq and Afghanistan, might contemplate military action yet again in the Middle East. On some level, senior U.S. policymakers may simply have been relieved that Israel struck quietly, resolved its security concerns, and spared the United States any further involvement. (Needless to say, an Israeli strike on Iran could well be very different.)

In the end, what may be most striking about the many accounts thus far released about Bush administration decisionmaking about the Syrian reactor is that they spend so little time discussing the obvious policy option—a demurral from Washington followed by a strike from Israel—that finally resolved the issue. Sherlock Holmes, after all, famously solved a case by listening for the dog that did not bark.

---

[68] Bush, 2010, p. 422.

[69] Rice, 2011, p. 708.

Conclusion

One should be wary of drawing overly neat policy lessons from history. Still, the specifics of the four preceding cases may be helpful to modern-day policymakers in both Israel and the United States, if only to see how some of their predecessors handled some cognate questions of notification of the United States. This study seeks to lift those individual strands out of the wider literature on the operations of 1956, 1967, 1981, and 2007 to provide additional perspective on the particular choices of earlier policymakers. Its principal policy value probably lies in the specifics of the cases, but a few additional closing reflections may also be of use.

As mentioned previously, several common factors can be seen at play throughout these four rather different historical episodes. These factors include the Israeli perception of the imminence of the threat that may be acted against; the (usually more detached) U.S. perception of that threat's danger to Israel; the importance of the U.S. interests implicated by potential Israeli military action; the nature of U.S. involvement in the broader Middle East at the time of potential Israeli military action, including the deployment of U.S. forces in the region; the overall state of play in U.S.-Israel relations, including other sources of bilateral tension or cooperation and recent episodes of strain or partnership; the relationship between the Israeli prime minister and the U.S. president, including both of their styles and ideologies; the nature of U.S. diplomatic efforts to resolve the problem; the lessons learned by both Israeli and U.S. decisionmakers about past episodes involving Israeli preemptive or preventive strikes; and the U.S. administration's overall view of Israel's relative strategic utility in the Middle East.

These factors run through the cases, providing some analytical framework for examining the different outcomes. The following paragraphs consider each of our four cases through the prism of these factors, in roughly the order listed above.

In Suez, Ben-Gurion saw a looming but not imminent threat in Nasser's Egypt, and he seized on French and (to a somewhat lesser degree) British outrage at Nasser to join up with the region's former rulers to launch a preventive war to try to cut Nasser down to size. The United States was not at all convinced that Israel needed to strike in 1956, and Eisenhower's anger was compounded by the damage that Suez—and the image of buccaneering imperial powers again indulging in gunboat diplomacy in the Third World—did to the Western position in the Cold War. Eisenhower came to conclude that Israel was often a source of regional instability, prone to lashing out, rather than a reliable partner. Ben-Gurion's longtime desire to woo the United States clashed with his obdurate personality and the sense of mission he felt about reducing the threat from Nasserite Egypt.

In 1967, mindful of the lessons of Suez, the more tractable and mild-mannered Eshkol made a pronounced and prolonged point of consultation with Washington, even as ordinary Israelis grew more and more alarmed that Nasser's increasingly belligerent behavior could endanger the Jewish state's very existence. Israel saw the threat in 1967 as imminent and immediate, and the waiting period—known in Hebrew as the *Hamtana*—remains a vivid and traumatic memory for Israelis who lived through it. The United States was more persuaded in 1967 that Israel faced an urgent threat than it had been in 1956, when Nasser's strength was growing but his behavior was not as directly and immediately menacing. As in 1956, the United States did not want to see its Cold War position in the Middle East undermined, but U.S. policymakers did not see possible Israeli action in 1967 as being as inimical to U.S. interests as the surprise of 1956 had been. U.S.-Israel relations in 1967 were a far cry from today's special relationship, but Johnson (unlike Eisenhower) had a marked gut-level warmth for the Jewish state, and the soft-spoken, humorous Eshkol was better at massaging foreign counterparts than the indomitable but irascible Ben-Gurion. Eshkol did his best to provide time for the United States to work through its playbook of steps to stave off conflict, including the ill-fated Red Sea Regatta. All these factors came together to produce an episode that was severely stressful for Israel's citizens but placed the emerging U.S.-Israel relationship under only manageable strain—and, indeed, helped paved the way for greater cooperation and closeness still.

In 1981, we see a rather different case. For starters, both the bombing of Osiraq and the attack on Syria's al-Kibar reactor were isolated raids rather than regional wars, which made for lower strategic stakes (though policymakers confronted with what is being sold as a one-off attack can never be sure it will not spark a wider conflagration). Of our four cases, only the Six-Day War seems to truly qualify as a *preemptive* strike rather than a *preventive* use of force; in 1956, 1981, and 2007, Israel may have faced enormous strategic risks down the line, but only in 1967 could it credibly be said to have been (to borrow an image from Michael Walzer) flinging up its hand to lessen the force of a blow that was already falling. The Reagan administration disliked the idea of a nuclear-armed Iraq but did not share Begin's sense of Holocaust-infused urgency. In the wake of the Iran hostage crisis, Washington saw Iran as its major bête noire in the region, not Iraq. The U.S.-Israel special relationship was well advanced by this point, with routinized and regularized contact on a broad range of diplomatic, military, and economic issues; significant foreign aid after the 1979 Egypt-Israel peace treaty; and a robust pro-Israel lobby with considerable clout on Capitol Hill. After decades of Labor Party prime ministers in Israel, Washington was dealing with the first Likud premier—and not always finding the experience easy. For all of Reagan's bonhomie, the sunny Californian and the dour, hawkish Begin were something less than a natural fit. The United States had not been deeply invested in efforts to deal with Iraq's nuclear program, but Reagan's team did take a friendly view of Israel's utility and reliability as a Western partner and thriving democracy. These factors did not obviate U.S. anger over being surprised by Begin's raid, but they did cushion it.

Finally, in 2007, Israel struck again at what it saw as a growing but not imminent threat. The United States, now embroiled in painful post-9/11 wars in Iraq and Afghanistan, was again unconvinced of the urgency—and concerned by the presence of some 140,000 U.S. troops in Iraq. The Bush administration had warm overall ties with Israel, granting both Olmert and his predecessor, Ariel Sharon, considerable post-9/11 license to move vigorously against terrorism from Hamas, Hizballah, and more secular Palestinian militias such as the Tanzim. Bush himself was strongly predisposed to strong action against terrorism and WMD—and, having invaded Iraq to topple Saddam Hussein, not in much of a position to denounce other democracies for failing to grasp the difference between preventive and preemptive strikes. The Bush administration—led by a president whose second inaugural famously set the "ultimate goal of ending tyranny in our world"—saw Israel, on balance, as a strategic asset and a democratic island in a sea of autocracy.[1] Olmert's decision to consult with Washington before striking, then, is only one of the factors behind the mild U.S. response to the raid on al-Kibar.

It is less clear how our list of factors would play out with a surprise Israeli raid on Iran. Cases drawn from the historical record sit uneasily alongside surmise about potential future events, of course, but a few tentative observations drawn from our list of key factors may be useful.

Israel and the United States seem to have a significant difference of opinion—lasting now throughout both Democratic and Republican administrations—about the imminence of the Iranian nuclear threat. Israeli attempts to persuade Republican and Democratic policymakers alike of the urgency have yet to produce a green (or even amber) light for the use of Israeli force. The strategic stakes for the United States in an Israeli strike on Iran would be massive. U.S. involvement in the Middle East is less today than it was at the height of the Iraq surge, but a war-weary American public may well balk at any developments that seem even to threaten renewed U.S. military involvement in the Middle East. Meanwhile, the region is roiling, with Arab politics shaken to their foundations by the revolutions in 2011 that toppled durable autocracies in Tunisia, Egypt, Libya, and Yemen and now shaken again by the counterrevolution launched in 2013 by the Egyptian military and the worsening Syrian civil war. U.S.-Israel relations remain deeply institutionalized and strong, particularly after President

---

1 George W. Bush, Second Inaugural Address, January 20, 2005.

Obama's successful March 2013 visit to Israel, but the Obama administration's first-term relationship with Israel was marred by strains over Israeli-Palestinian peace and ongoing Israeli settlement activity. Obama and Netanyahu, a Democrat and a Likudnik, have managed to work out a largely cordial working relationship, but some administration officials reportedly think Netanyahu has dragged his heels on peace efforts and made scant effort to hide his preference for former Massachusetts governor Mitt Romney in the 2012 election campaign. The United States, over several administrations, has tried to pressure Iran to abandon its nuclear-arms ambitions through sanctions and diplomacy—backed up by warnings that (as Obama put it in Jerusalem) "all options are on the table." The United States, the president added, "will do what we must to prevent a nuclear-armed Iran."[2] The United States and Israel have also gone through several episodes in which Israel has hinted that action against Iran may be becoming more imminent, all of which have ended without Israeli strikes. Taken together, our now-familiar list of factors suggest that whatever a surprise Israeli strike on Iran might bring, it would probably not be greeted with the relative calm of the U.S. response in 2007 or even 1981.

As the foregoing suggests, Israel's strategic thinking has to grapple with a central tension. As a small, embattled state in a hostile neighborhood, Israeli leaders often believe they dare not rely on outsiders to resolve their security dilemmas—but Israeli leaders also must work to safeguard their greatest strategic asset: Israel's special relationship with the world's most powerful country.

Israeli leaders have sometimes wobbled on this balance beam, but they have rarely fallen off it. After a brief flirtation with nonalignment early in Israel's history, its leaders have always sought great-power backing. David Ben-Gurion, Israel's first prime minister, made the support of a great power the central pillar of the young state's foreign policy. While Ben-Gurion was willing to join with Britain to try to topple Nasser, Israel's founding father always aspired to an alliance with the

---

[2] Barack Obama, Transcript of Obama's speech in Israel, *New York Times*, March 21, 2013.

United States. As Shabtai Teveth, Ben-Gurion's biographer, puts it, "What he really wanted was America in his corner."[3]

The special relationship between the United States and the state of Israel remains a mighty wellspring of Israeli security and a massive strategic asset, providing Israel with state-of-the-art weaponry, vital military funding, military-to-military bonds, foreign aid, intelligence cooperation, and political cover in the UN Security Council and elsewhere. But if Israel cannot live without the alliance, it does not always live entirely easily with it. Israel's understanding of its security requirements is not necessarily the same as America's, and the policy judgments of one ally's leadership are not necessarily the same as that of the other.

Today's U.S.-Israel special relationship is a very far cry from the tense and brittle exchanges of the Eisenhower era. The four cases we have examined, from Suez to al-Kibar, trace the evolution of the alliance—from the mistrust, skepticism, and even outright hostility of the mid-1950s to the capacious and comprehensive interwoven relationship so familiar to us today. That has shifted the calculus of Israeli leaders over the decades in trying to weigh the wisdom of consulting with the United States before using force preemptively or preventively, particularly in the context of a burgeoning nuclear threat.

It is always risky to draw direct policy conclusions from even recent history. Perhaps the most obvious conclusion for U.S. policy is one that seems from multiple open sources to have already occurred: The United States seems overwhelmingly likely to have insisted on significant advance notification from Israel before any military strike against Iran's nuclear facilities. But even beyond this point, our four cases do suggest a few conclusions worth considering for the years ahead, for policymakers in both Washington and Jerusalem.

---

[3] Bass, 2003, p. 5.

Superpowers Hate Surprises

Kant has placed ingratitude among the vices that are "the essence of vileness and wickedness." Even if that puts it too strongly, superpowers do not appreciate being startled by small and vulnerable states that they have done much to support.

Eisenhower and Dulles remain exemplars of this point. Some of Eisenhower's fury over Suez may have been personal; he was grappling with health issues, reaching the height of his reelection campaign, and dealing with a secretary of state who had been laid low with terminal cancer during the crisis. But the more Eisenhower learned about Suez, the angrier he got about one of the Cold War's biggest surprises. As the historian David Schoenbaum archly puts it, "as the picture gradually cleared, [Eisenhower] only became more furious at what he clearly viewed as a marriage of stupidity and betrayal."[4] Ben-Gurion had hardly assumed that Washington would applaud the campaign to topple Nasser, but he does not seem to have counted on an eruption of molten lava either.

On a much smaller scale, the Reagan administration was also distinctly put out by Israel's 1981 surprise raid on Osiraq. While Reagan's aides have disagreed about the extent of Reagan's annoyance or approbation, the U.S.-Israel strains were clear to see. Even before the raid, the United States and Israel had been at loggerheads over the AWACS controversy, and whatever the president's own emotional attachments to the Jewish state, Secretary of State Haig and Secretary of Defense Weinberger both found Israeli unilateralism a hindrance in their regional efforts. As in 1956, these troubling trend lines were exacerbated in 1981 by the element of surprise. And many of Reagan's advisers remembered their dislike for Begin's Osiraq tactics a year later when the IDF rolled into Beirut.

Israeli leaders may sometimes conclude that they have no choice but to keep Washington in the dark. But they should understand that U.S. policymakers may well bristle. If Israeli leaders may sometimes insist on making their own decisions about their own security without

---

[4] Schoenbaum, 1993, p. 115.

input from the United States, U.S. leaders will always prefer making their own security decisions without jolts from smaller powers that enjoy American assistance.

U.S. Irritation Is Bounded by the Evolution of the Special Relationship

Perhaps what is most striking about Eisenhower and Dulles's response to the Suez shock is the sense that most everything was on the table. Even UN sanctions were not beyond the pale. In early 2009, by contrast, the departing Bush administration found itself in a significant tussle with the Olmert government in Israel for simply abstaining on a Security Council resolution about the short, sharp mini-war in Gaza. (Secretary of State Rice, who played a significant role in negotiating the resolution, had wanted the administration to vote for it but was ordered to abstain by President Bush, who had just received an earful from a fuming Olmert.) But if in 1956 the entire relationship between the United States and Israel seemed to hang in the balance, in 2009 it was clear to both sides that the resolution was a relative hiccup and that the wider, professionalized interactions of the alliance would continue substantially undisturbed.

Similarly, the breadth and depth of the special relationship made it easier for Olmert in 2007 to sound out the Bush administration over the Kibar reactor—even to hope that hawkish elements inside the administration might carry the day over Israel's request for a U.S. bombing run to take out the Syrian facility. (It takes very little imagination to gather that Vice President Cheney, senior NSC staffer Elliott Abrams, and others later quietly applauded the Israeli raid.) But beyond the individual policy views of the key decisionmakers, the structural constraints of the fully articulated special relationship now provide something of a shock absorber—and while it would be a mistake to assume that the U.S.-Israel relationship can handle any jolt, no matter how sharp, it will also tend to exert significant pressure in the direction of continuity after even a major surprise.

Israeli Action Is Bounded by the Depth of U.S. Interests

The Bush administration's sanguine response to the 2007 Kibar raid might have been rather different if the attack had somehow triggered a regional cascade that brought in Iran, Syria's closest regional partner, and seriously set back the U.S. counterinsurgency campaign gaining traction in Iraq. At the risk of being tautological, Israel got away with it because Israel got away with it. Had the raid triggered a major regional crisis, Israel's attack might have spurred a somewhat different U.S. response. In 1967, Israel was pressured to endure weeks of ultimately fruitless diplomacy because the Johnson administration was wary of unleashing a chain of events that could have compelled it to intervene in the Middle East even as the war in Vietnam was escalating. Hypothetically, if Israel were to discover that a hostile hermit of a nation without economic significance, political friends, or strategic value was working on a tiny, uninhabited island to produce a nuclear reactor, Israel would have considerably more ability to choose military action without risking U.S. goodwill. But the more embedded a potential target is in core U.S. security interests, the less leeway Israel is likely to have—and the riskier it will be to surprise a global power with interests and even personnel in harm's way.

Consultation Does Not Guarantee U.S. Assent, but It Can Limit U.S. Anger

In both 1967 and 2007, Israel's leadership consulted extensively with Washington before resorting to the use of force. In 1967, Eshkol, Eban, and other Israeli interlocutors sought an American green light for the use of force, only to be led into the back alleys of lengthy discussions about the so-called Red Sea Regatta—Johnson's proposed (and never overwhelmingly plausible) alternative to reopen the Straits of Tiran short of the use of outright force. Israel helped its postwar position by spending considerable prewar time and energy exhausting the proposed U.S. option. Eshkol never got a formal green light, and while LBJ's carefully repeated mantra—"Israel will not be alone unless it goes alone"—may have been vague, it hardly encouraged Israeli preemption. But Eshkol could at least reassure himself that the Johnson administration would not be surprised by Israel's strikes. The Johnson administration's response to the outbreak of war in 1967 might well have been somewhat different if the war had not been such a triumph for the IDF, but even then it seems likely that Eshkol's investment in consultations would have paid dividends in American understanding. While some of Johnson's senior advisers had at best mixed emotions over the Israeli decision to strike, they all had to at least concede that Eshkol's ultimate decision, after weeks of tense discussions, was far from surprising.

A somewhat similar dynamic can be observed in 2007, when Ehud Olmert made a far more explicit request—not for U.S. sympathy but for U.S. bombing runs to take out the Kibar reactor in Syria. The Bush administration ultimately overruled Vice President Cheney and a handful of other hawks who wanted to launch a U.S. raid and decided instead to propose a diplomatic route, taking intelligence on the Syrian reactor to the IAEA and bringing the issue before the UN Security Council. Olmert may well have regarded this as somewhat akin to the Red Sea Regatta—a time-buying diversion that did not address the underlying and urgent threat to Israeli security. But like Eshkol (and unlike Ben-Gurion and Begin), Olmert took the time to ventilate the issue in considerable detail with the most senior U.S. officials. Bush administration officials immediately grasped that they were facing the prospect of a unilateral Israeli raid inside Syria if they chose not to act themselves. By declining to launch a U.S. raid, the Bush administration understood that it might well have to manage the consequences of an Israeli one. But at least one of those consequences would not be calming a surprised superpower already tied down in two wars in the greater Middle East. In the end, Olmert did not take the Bush team's advice, but the Bush team was given ample strategic warning that Israeli action was highly likely.

Different U.S. Agencies Will Often Differ

The old bureaucratic saying holds that where you stand depends on where you sit. This will inevitably continue to be the case when it comes to future cases of Israeli preemption or preventive war. Defense Department officials will be deeply concerned about the consequences for U.S. troops already in harm's way in the region, while State Department officials will worry about the impact of Israeli action on regional relationships that are invariably under structural strain. This will always complicate communication between capitals and tempt even U.S. allies to play one part of America's interagency system against another. In 1967, for instance, Israel got far closer to the green light it so ardently sought in the response from Pentagon officials, who had analyzed the balance of forces between Israel and Egypt and were far more confident in the IDF's ability to prevail than were most of the deeply alarmed Israeli citizenry. In 2007, however, Olmert ran into a bureaucratic brick wall in the form of National Security Adviser Hadley, Defense Secretary Gates, and Secretary of State Rice, none of whom were enthusiastic about the prospect of another U.S. military campaign in the Muslim world. Vice President Cheney found himself isolated in the interagency, making it easier for President Bush to side with his senior secretaries.

In the real world, U.S. policymakers should probably expect Israeli governments to probe for signs of internal division or disagreement—and Israeli policymakers should probably expect American governments to bristle at any such probing. (Of course, Israeli agencies will also disagree.)

The View from Washington Is Different than the View from Jerusalem

It takes nothing away from the urgency of Israel's national security imperatives to recall that small powers do not invariably seek out the big picture. An Israeli prime minister will, understandably, be most concerned with saving the Jewish state from the threat of nuclear annihilation; an American president will wholeheartedly share that vital goal but will also pull back the lens to consider the wider state of world politics and the global nonproliferation regime. American presidents will have to consider not only U.S. security writ large, including the deployments of U.S. troops and the safety of U.S. diplomatic facilities and the homeland itself, but also the growth of the U.S. economy, the stability of U.S. sources of energy, and the success of U.S. grand strategy, among other calculations befitting a global power with global interests.

Moreover, Israel's deadline will always be earlier than America's, and Israel's tolerance for uncertainty and risk while diplomacy unfolds will always be lower than America's. The United States' geography and vast arsenal gives it more of a cushion on timing and defers the anxiety that military action could come too late.

While the United States does not face the same types of direct, short-term, existential threats to its security that Israel does, U.S. security is still directly embroiled in the torments of the Middle East. The United States has an ambitious array of goals in the Middle East today, from preventing Iran from acquiring nuclear weapons to dismantling terrorist networks that threaten U.S. citizens to managing the aftermath of the 2011 Arab revolutions. Israeli leaders have usually shared the broader goals of U.S. strategy but have sometimes given short-term precedence to what they viewed as pressing imperatives vital to the survival of the Jewish state.

The different perspectives were perhaps most evident during the Suez crisis. Ben-Gurion was determined to overthrow (or at least weaken) Israel's nemesis, while Eisenhower was enraged by what he considered a galling display of gunboat diplomacy that would cost the West precious ground in the Cold War.

Israel Should Think Hard Before Deciding to Beg for Forgiveness Rather than Ask for Permission

Looking back on these cases, few Israeli decisionmakers will find overpowering reason for optimism about securing U.S. blessing for a strike soon on Iran's nuclear facilities. When Eshkol and Olmert beseeched Washington for a green light to strike, they did not get it—and even though the act of consultation paid its own dividends, it did not actually shift the stated U.S. preference for avoiding the use of force. As such, the Osiraq case may prove a tempting precedent to Israeli decisionmakers. After all, while the Reagan administration did express real dislike for Israel's actions, Begin's raid did not cause a strategic rupture in the special relationship. Some Israelis have long argued that results will speak for themselves and that a U.S. administration that seems publicly irritated over an Israeli strike may prove privately grateful.

In other words, Israeli leaders may choose to ask the United States for forgiveness for striking Iran rather than pleading for permission to do so. Both the Bush administration and the Obama administration seem to have urged Israel not to attack Iranian nuclear facilities, which may mean that Israeli decisionmakers will be more tempted than in past cases to avoid seeking an American green light that may never be forthcoming.

But Osiraq is an inexact analogy here, and Israeli officials should not assume that the United States will simply experience mild heartburn over a raid on Iran and then get over it. After a raid on Iran, Israeli leaders may think their boldness has opened a rare window of opportunity to weaken Iran's strategic position—and even perhaps the mullahs' grip on power. But U.S. officials may take a far more grim view, believing that what has been opened is not a shiny gift but Pandora's box.

In any event, top-tier U.S. national security priorities would probably be directly engaged in an Israeli strike on Iran. U.S. forces and assets in the region could be targeted in Iranian reprisals. Global oil prices would surely be jolted. Hizballah could be unleashed to bombard Israeli cities or stage terrorist strikes abroad, opening another conflict between Israel and Lebanon. Other U.S. partners such as Saudi Arabia could also find themselves drawn in.

Given these and other risks to U.S. interests, Israeli leaders who surprised Washington by striking Iran might find themselves begging for a forgiveness that is not particularly likely to be forthcoming. U.S. policymakers might well see their Israeli counterparts as reckless, immature, and cocky; Israeli policymakers, in turn, may see their U.S. counterparts as shortsighted, sanctimonious, and hidebound. The stakes for the U.S.-Israel special relationship in any Iran strike are likely to be far higher than they were in 1981 or 2007—and the consequences may be more lasting too.

Presidents Like Consulters Better than Confronters

The United States and Israel have held close consultations for several years over Iran's nuclear program. That is as it should be. If this study has a bottom line, it is simply this: For an Israeli prime minister contemplating the use of preemptive or preventive force, the risks of informing Washington have historically tended to be lower than the risks of blindsiding it. Israel has no more precious outside security asset than the support and friendship of the United States—period, full stop. Like all valuable relationships, the U.S.-Israel alliance requires ongoing effort—even hard work. Consultation over even the most difficult security matters occurs within the framework of the special relationship; confrontation is by definition a high-stakes exercise that simply hopes that Washington will applaud or accept a *fait accompli*. If a relationship is about dialogue, confrontation takes a step outside of it. That is not to say that a confrontation equals a rupture. It is simply to note that Eshkol and Olmert were able to avoid major shocks to the U.S.-Israel special relationship and still tackle major threats to Israeli security. The U.S.-Israel alliance has come a long day from Eisenhower's eruption over Suez or the Reagan administration's willingness to take the Osiraq raid to the UN Security Council. But in the months and years ahead, any American president is going to want to be consulted about matters that directly affect the vital interests of the United States—and any Israeli prime minister who chooses not to do so will have to weigh the security gains of the attack in question against the security losses from running the risk of a painful episode between an embattled regional pariah and its most powerful friend.

References

Abrams, Elliott, "Bombing the Syrian Reactor: The Untold Story," *Commentary*, February 2013.

Allen, Richard V., "Reagan's Secure Line," *New York Times*, June 7, 2010.

"Attack—and Fallout," *TIME*, June 22, 1981, Vol. 117, No. 25, p. 28.

Bar-Zohar, Michael, *Shimon Peres: The Biography*, New York: Random House, 2007.

Bass, Warren, *Support Any Friend: Kennedy's Middle East and the Making of the U.S.-Israeli Alliance*, New York: Oxford University Press, 2003.

Begin, Menachem, open letter to American Jews and Christians, Menachem Begin Heritage Center archives, June 12, 1981.

Brands, Hal, and David Palkki, "Saddam, Israel, and the Bomb: Nuclear Alarmism Justified?" *International Security*, Vol. 36, No. 1, 2011, pp. 133–166.

Brecher, Michael, *Decisions in Israel's Foreign Policy*, New Haven: Yale University Press, 1975.

Bush, George W., Second Inaugural Address, January 20, 2005. As of November 4, 2013:
http://www.npr.org/templates/story/story.php?storyId=4460172.

———, statement at press conference, September 20, 2007. As of November 4, 2013:
http://georgewbush-whitehouse.archives.gov/news/releases/2007/09/20070920-2.html

———, *Decision Points*, New York: Crown, 2010.

Cannon, Lou, *President Reagan: The Role of a Lifetime*, New York: PublicAffairs, 2000.

Case Bryant, Christa, "Obama-Netanyahu Tensions: Not as Bad as 5 Other US-Israel Low Points," *Christian Science Monitor*, undated. As of November 4, 2013:
www.csmonitor.com/World/Middle-East/2012/0927/Obama-Netanyahu-tensions-Not-as-bad-as-5-other-US-Israel-low-points/1981-Israeli-strike-on-Osirak-nuclear-reactor-in-Iraq

Cheney, Dick, with Liz Cheney, *In My Time: A Personal and Political Memoir*, New York: Threshold, 2011.

Cohen, Avner, *Israel and the Bomb*, New York: Columbia University Press, 1998.

Eban, Abba, *Personal Witness: Israel Through My Eyes*, New York: Putnam, 1992.

Haig, Alexander, letter to Senate Foreign Relations Committee Chairman Charles Percy, Menachem Begin Heritage Center archives, June 11, 1981.

Haig, Alexander, *Caveat: Realism, Reagan, and Foreign Policy*, New York: Macmillan, 1984.

Hayden, Michael, "Correcting the Record About that Syrian Nuclear Reactor," *Washington Post*, September 22, 2011.

Hersh, Seymour, *The Samson Option: Israel's Nuclear Arsenal and American Foreign Policy*, New York: Random House, 1991.

Herzog, Chaim, (updated by Shlomo Gazit), *The Arab-Israeli Wars: War and Peace in the Middle East from the 1948 War of Independence to the Present*, New York: Vintage, 2005.

Israel Government Press Office, "Press Conference with Prime Minister Begin, Chief-of-Staff Eitan, Air Force Commander Ivri, and Director of Military Intelligence Saguy," *Press Bulletin*, June 9, 1981.

Jessup, Peter, interview with Samuel Lewis, Association for Diplomatic Studies and Training, Foreign Affairs Oral History Project, August 9, 1998.

Johnson, Lyndon B., *The Vantage Point: Perspectives of the Presidency, 1963–69*, New York: Holt, Rinehart, & Winston, 1971.

Kahl, Colin, "Before Attacking Iran, Israel Should Learn from Its 1981 Strike on Iraq," *The Washington Post*, March 2, 2012. As of November 4, 2013:
http://articles.washingtonpost.com/2012-03-02/opinions/35450430_1_nuclear-weapons-israeli-strike-tuwaitha

Karpin, Michael, *The Bomb in the Basement: How Israel Went Nuclear and What That Means for the World*, New York: Simon & Schuster, 2007.

Kyle, Keith, *Suez*, New York: St. Martin's, 1991.

Lewis, Samuel, interview with the author, Dec. 27, 2012.

"Long Shadow of the Reactor," *TIME*, Vol. 118, No. 1, July 6, 1981, p. 35.

Makovsky, David, "The Silent Strike," *The New Yorker*, September 17, 2012.

———, discussion with author, January 16, 2013.

Mueller, Karl, Jasen J. Castillo, Forrest E. Morgan, Negeen Pegahi, and Brian Rosen, *Striking First: Preemptive and Preventive Attack in U.S. National Security Policy*, Santa Monica, Calif.: RAND Corporation, MG-403-AF, 2006. As of November 4, 2013:
http://www.rand.org/pubs/monographs/MG403.html

Nuclear Threat Initiative, "Country Profile: Israel," August 2013. As of November 4, 2013:
http://www.nti.org/country-profiles/israel/nuclear/

Obama, Barack, Transcript of Obama's speech in Israel, *New York Times*, March 21, 2013. As of November 4, 2013:
http://www.nytimes.com/2013/03/22/world/middleeast/transcript-of-obamas-speech-in-israel.html?pagewanted=all&_r=0

Oren, Michael, *Six Days of War: June 1967 and the Making of the Modern Middle East*, New York: Oxford University Press, 2002.

———, *Power, Faith, and Fantasy: America in the Middle East, 1776 to the Present*, New York: Norton, 2007.

Parker, Richard B., *The Politics of Miscalculation in the Middle East*, Bloomington: Indiana University Press, 1993.

Peres, Shimon, *Battling for Peace: A Memoir*, New York: Random House, 1995.

Pollack, Kenneth, *The Threatening Storm: The Case for Invading Iraq*, New York: Random House, 2002.

Quandt, William B., *Decade of Decisions: American Policy Toward the Arab-Israeli Conflict, 1967–1976*, Berkeley: University of California Press, 1977.

Reagan, Ronald, statement at press conference, Menachem Begin Heritage Center archives, June 17, 1981.

"Reagan Diaries: Carter Failed to Tell Reagan About Osirak," History News Network, May 28, 2007.

Reich, Bernard, ed., *An Historical Encyclopedia of the Arab-Israeli Conflict*, Westport, Conn.: Greenwood, 1996.

Rice, Condoleezza, *No Higher Honor: A Memoir of My Years in Washington*, New York: Crown, 2011.

Rubin, Trudy, "Habib Pushing Syrian Missile Crisis onto Wider Stage," *Christian Science Monitor*, May 26, 1981.

Sachar, Howard, *A History of Israel, Volume II: From the Aftermath of the Yom Kippur War*, New York: Oxford University Press, 1987.

Sanger, David, *The Inheritance: The World Obama Confronts and the Challenges to American Power*, New York: Three Rivers, 2009.

Schoenbaum, David, *The United States and the State of Israel*, New York: Oxford University Press, 1993.

Spiegel, Steven, *The Other Arab-Israeli Conflict: Making America's Middle East Policy from Truman to Reagan*, Chicago: University of Chicago Press, 1986.

United Nations Security Council, Resolution 487, June 19, 1981. As of November 4, 2013:
http://www.yale.edu/lawweb/avalon/un/un487.htm

Walzer, Michael, *Just and Unjust Wars: A Moral Argument With Historical Illustrations*, New York: Basic, 2006.

Weinberger, Caspar, *Fighting for Peace: Seven Critical Years in the Pentagon*, New York: Warner, 1991.

Woodward, Bob, "In Cheney's Memoir, It's Clear Iraq's Lessons Didn't Sink In," *Washington Post*, September 11, 2011.